STUDENT LECTURE NOTEBOOK

ESSENTIALS OF GEOLOGY

NINTH EDITION

Frederick K. Lutgens
Edward J. Tarbuck

Upper Saddle River, NJ 07458

Editor-in-Chief: Dan Kaveney
Executive Editor: Patrick Lynch
Project Manager: Dorothy Marrero
Executive Managing Editor: Kathleen Schiaparelli
Assistant Managing Editor: Becca Richter
Production Editor: Gina M. Cheselka
Supplement Cover Manager: Paul Gourhan
Supplement Cover Designer: Joanne Alexandris
Manufacturing Buyer: Alexis Heydt-Long
Cover Photo Credit: Bill Hatcher/National Geographic Image Collection

© 2006 Pearson Education, Inc.
Pearson Prentice Hall
Pearson Education, Inc.
Upper Saddle River, NJ 07458

All rights reserved. No part of this book may be reproduced in any form or by any means, without permission in writing from the publisher.

Pearson Prentice Hall™ is a trademark of Pearson Education, Inc.

The author and publisher of this book have used their best efforts in preparing this book. These efforts include the development, research, and testing of the theories and programs to determine their effectiveness. The author and publisher make no warranty of any kind, expressed or implied, with regard to these programs or the documentation contained in this book. The author and publisher shall not be liable in any event for incidental or consequential damages in connection with, or arising out of, the furnishing, performance, or use of these programs.

> **This work is protected by United States copyright laws and is provided solely for teaching courses and assessing student learning. Dissemination or sale of any part of this work (including on the World Wide Web) will destroy the integrity of the work and is not permitted. The work and materials from it should never be made available except by instructors using the accompanying text in their classes. All recipients of this work are expected to abide by these restrictions and to honor the intended pedagogical purposes and the needs of other instructors who rely on these materials.**

Printed in the United States of America

10 9 8 7 6 5 4 3 2 1

ISBN 0-13-192743-4

Pearson Education Ltd., *London*
Pearson Education Australia Pty. Ltd., *Sydney*
Pearson Education Singapore, Pte. Ltd.
Pearson Education North Asia Ltd., *Hong Kong*
Pearson Education Canada, Inc., *Toronto*
Pearson Educación de Mexico, S.A. de C.V.
Pearson Education—Japan, *Tokyo*
Pearson Education Malaysia, Pte. Ltd.

Contents

To the Student iv
CHAPTER 1   An Introduction to Geology 1-1
CHAPTER 2   Minerals: Building Blocks of Rocks 2-1
CHAPTER 3   Igneous Rocks 3-1
CHAPTER 4   Volcanoes and Other Igneous Activity 4-1
CHAPTER 5   Weathering and Soils 5-1
CHAPTER 6   Sedimentary Rocks 6-1
CHAPTER 7   Metamorphic Rocks 7-1
CHAPTER 8   Mass Wasting: The Work of Gravity 8-1
CHAPTER 9   Running Water 9-1
CHAPTER 10 Groundwater 10-1
CHAPTER 11 Glaciers and Glaciation 11-1
CHAPTER 12 Deserts and Wind 12-1
CHAPTER 13 Shorelines 13-1
CHAPTER 14 Earthquakes and Earth's Interior 14-1
CHAPTER 15 Plate Tectonics: A Scientific Theory Unfolds 15-1
CHAPTER 16 Origin and Evolution of the Ocean Floor 16-1
CHAPTER 17 Crustal Deformation and Mountain Building 17-1
CHAPTER 18 Geologic Time 18-1
CHAPTER 19 Earth History: A Brief Summary 19-1
Appendices A-1

To the Student

This *Student Lecture Notebook* is designed to help you do your best in this physical geology course.

Key images from the textbook and every illustration from the Instructor's Transparency Set are reproduced in this notebook. Because you won't have to redraw the art in class, you can focus your attention on the lecture, annotate the art, and take your notes in this book.

Leave all your notes together or remove them for integration into a binder with other course materials.

CHAPTER 1 An Introduction to Geology

NOTES:

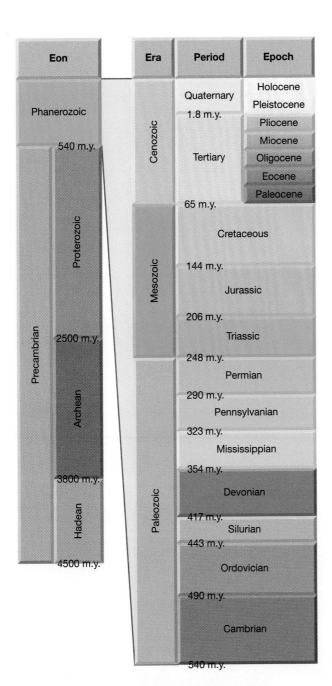

Figure 1.4 **Small geologic time scale.**

1-2   Chapter 1

NOTES:

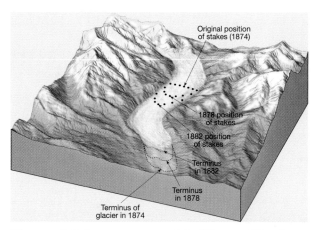

Figure 1.A **Ice movement in Rhone Glacier.**

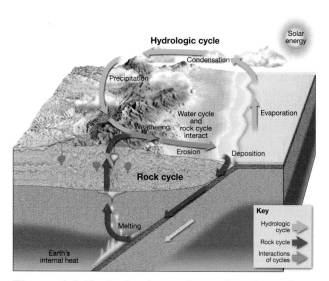

Figure 1.8 **Hydrologic cycle and rock cycle.**

An Introduction to Geology 1-3

NOTES:

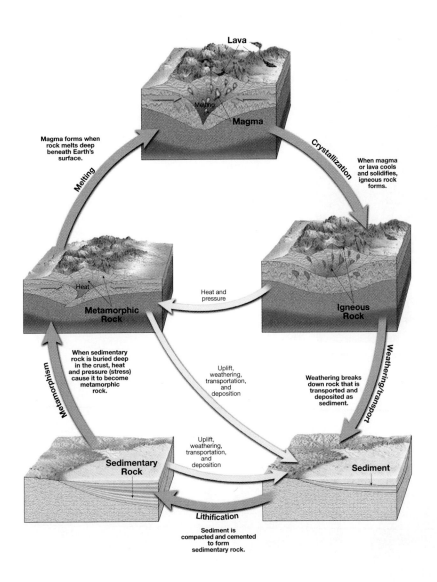

Figure 1.11 Rock cycle.

1-4 Chapter 1

Figure 1.12 Original solar nebula condenses into the solar system.

NOTES:

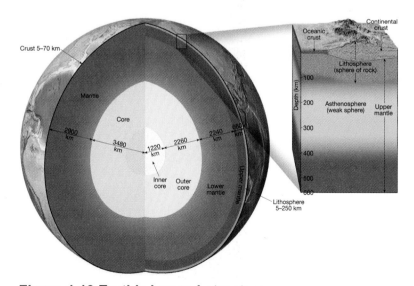

Figure 1.13 Earth's layered structure.

An Introduction to Geology   1-5

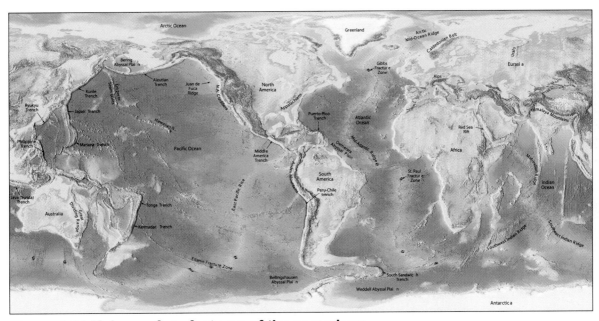

Figure 1.14 **Major surface features of the geosphere.**

NOTES:

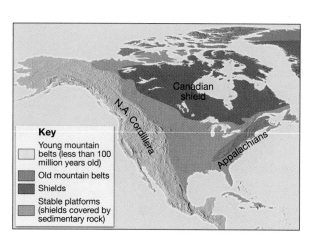

Figure 1.15 Major features of North America.

An Introduction to Geology 1-7

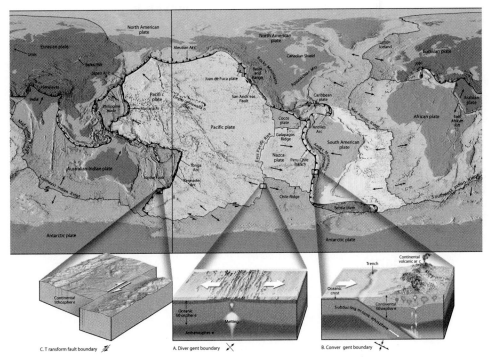

Figure 1.16 Plate map.

NOTES:

NOTES:

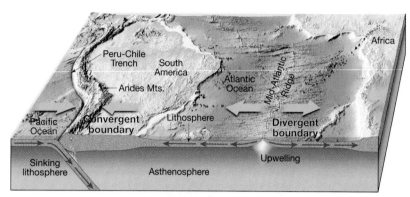

Figure 1.17 **Convergent boundaries.**

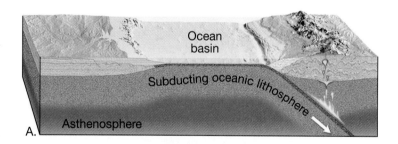

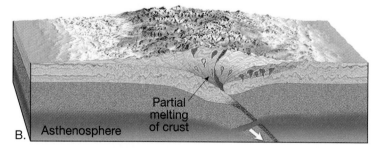

Figure 1.18 **Collision of India with Eurasian plate producing Himalaya Mountains.**

CHAPTER 2   Minerals: Building Blocks of Rocks

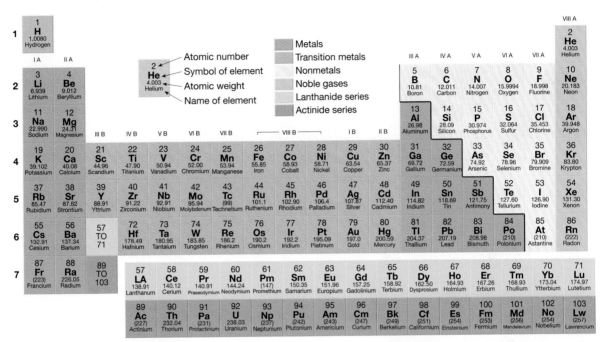

Figure 2.4 Periodic table.

NOTES:

Chapter 2

NOTES:

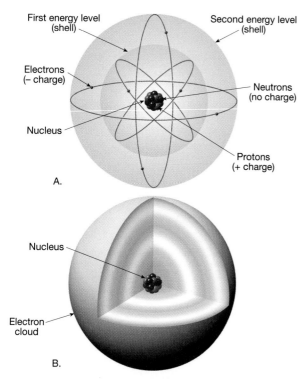

Figure 2.5 Two models of the atom.

Electron Dot Diagrams for Some Representative Elements							
I	II	III	IV	V	VI	VII	VIII
H·							He:
Li·	·Be·	·B·	·C·	·N·	:Ö·	:F·	:Ne:
Na·	·Mg·	·Al·	·Si·	·P·	:S·	:Cl·	:Ar:
K·	·Ca·	·Ga·	·Ge·	·As·	:Se·	:Br·	:Kr:

Figure 2.6 Electronic dot diagram.

Minerals: Building Blocks of Rocks 2-3

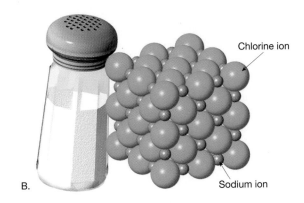

Figure 2.7 Chlorine, sodium with salt shaker.

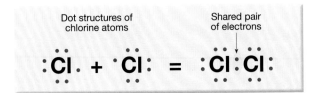

Figure 2.8 Dot structure of chlorine atoms.

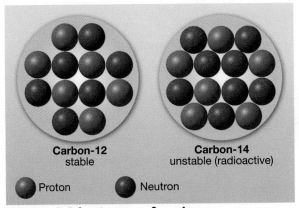

Figure 2.9 Isotopes of carbon.

NOTES:

NOTES:

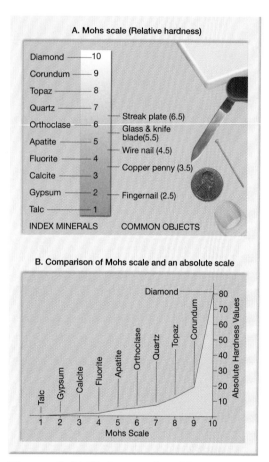

Figure 2.11 **Mohs scale of hardness.**

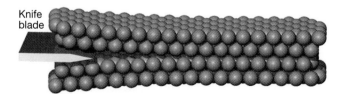

Figure 2.12 **Knife blade.**

Minerals: Building Blocks of Rocks 2-5

NOTES:

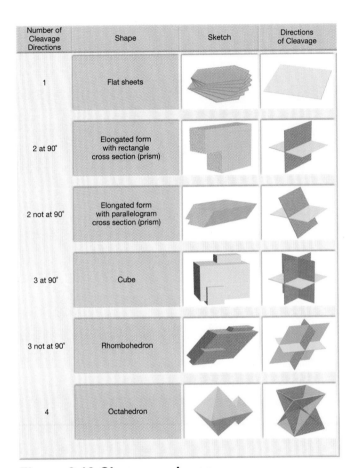

Figure 2.13 **Cleavage planes.**

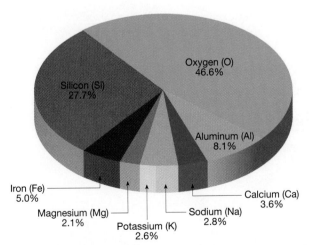

Figure 2.15 **Eight most abundant elements in Earth's crust.**

NOTES:

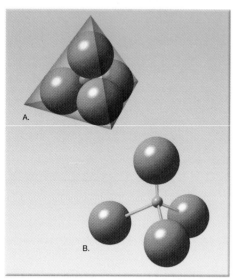

Figure 2.16 Two ball and stick models of silicon-oxygen tetrahedron.

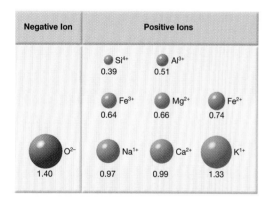

Figure 2.17 Sizes and charges of eight common elements of Earth.

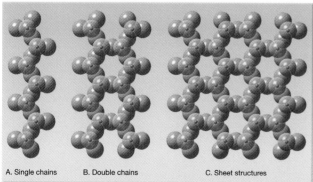

Figure 2.18 Three types of silicate structures: single, double, and sheet.

NOTES:

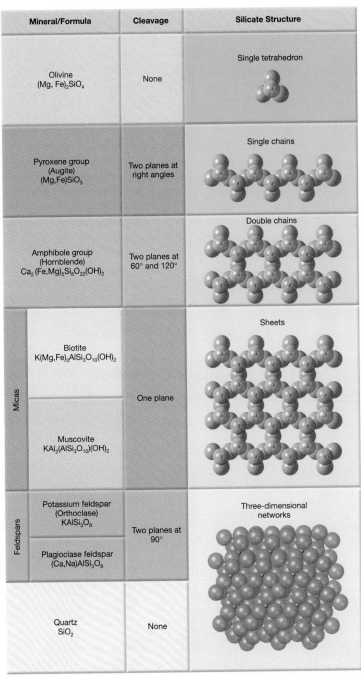

Figure 2.19 Common silicate minerals.

CHAPTER 3 Igneous Rocks

NOTES:

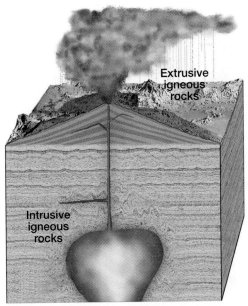

Figure 3.3 **Igneous rock textures.**

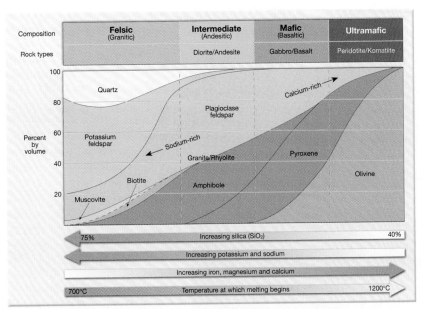

Figure 3.6 **Mineralogy of common igneous rocks.**

NOTES:

Chemical Composition		Granitic (Felsic)	Andesitic (Intermediate)	Basaltic (Mafic)	Ultramafic
Dominant Minerals		Quartz Potassium feldspar Sodium-rich plagioclase feldspar	Amphibole Sodium- and calcium-rich plagioclase feldspar	Pyroxene Calcium-rich plagioclase feldspar	Olivine Pyroxene
Accessory Minerals		Amphibole Muscovite Biotite	Pyroxene Biotite	Amphibole Olivine	Calcium-rich plagioclase feldspar
TEXTURE	Phaneritic (coarse-grained)	**Granite**	**Diorite**	**Gabbro**	**Peridotite**
	Aphanitic (fine-grained)	**Rhyolite**	**Andesite**	**Basalt**	**Komatiite** (rare)
	Porphyritic	"Porphyritic" precedes any of the above names whenever there are appreciable phenocrysts			Uncommon
	Glassy	**Obsidian** (compact glass) **Pumice** (frothy glass)			
	Pyroclastic (fragmental)	**Tuff** (fragments less than 2 mm) **Volcanic Breccia** (fragments greater than 2 mm)			
Rock Color (based on % of dark minerals)		0% to 25%	25% to 45%	45% to 85%	85% to 100%

Figure 3.7 Classification of igneous rocks.

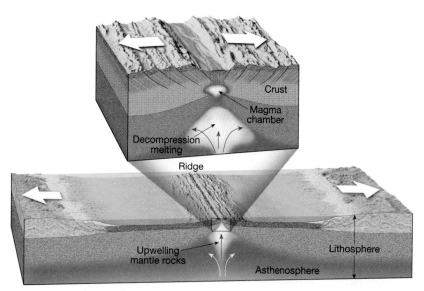

Figure 3.14 Drop in confining pressure triggers melting.

Igneous Rocks 3-3

NOTES:

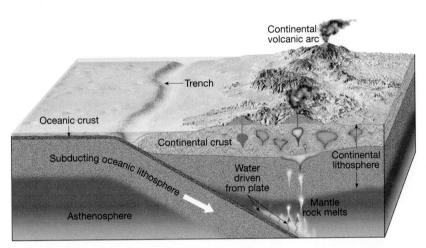

Figure 3.15 Oceanic plate descends into mantle, driving water and other volatiles from subducting crustal rock.

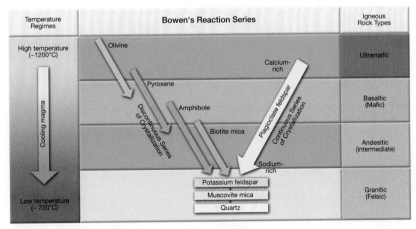

Figure 3.16 Bowen's reaction series.

3-4   Chapter 3

NOTES:

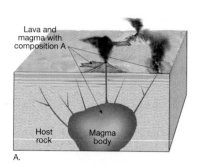
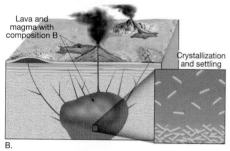
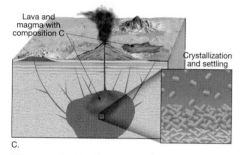

Figure 3.17 **How magma evolves.**

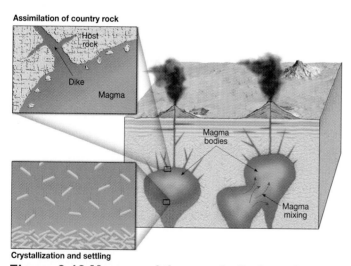

Figure 3.18 **Magma mixing, assimilation of country rock, and crystal settling.**

Igneous Rocks 3-5

NOTES:

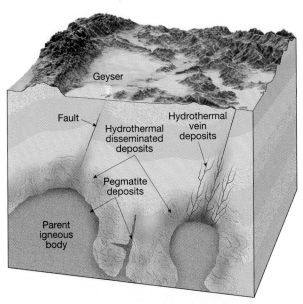

Figure 3.20 Relationship of igneous body with pegmatite and hydrothermal deposits.

CHAPTER 4   Volcanoes and Other Igneous Activity

NOTES:

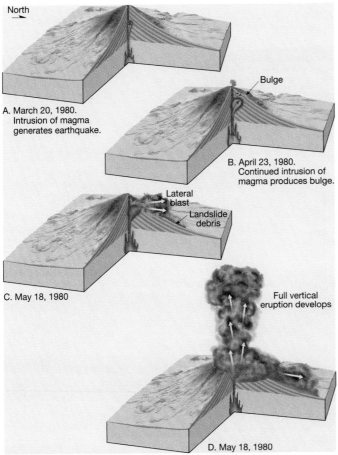

Figure 4.A Eruption of Mount St. Helens.

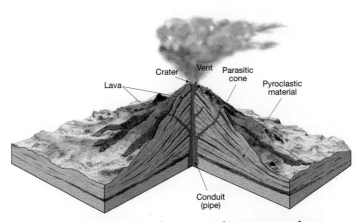

Figure 4.7 Anatomy of composite cone volcano.

NOTES:

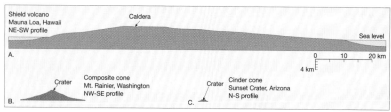

Figure 4.9 Profiles of volcanic landforms.

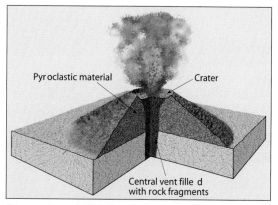

Figure 4.11 SP Crater, a cinder cone.

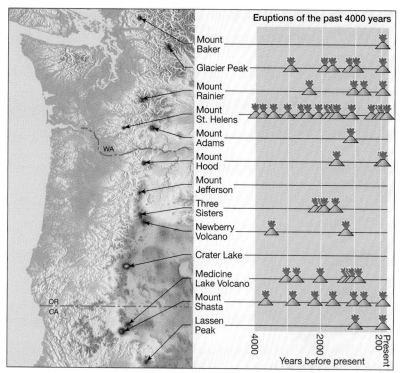

Figure 4.14 13 major volcanoes in Cascade Range.

Volcanoes and Other Igneous Activity 4-3

NOTES:

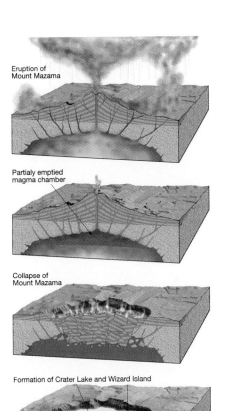

Figure 4.18 **Sequence of events that formed Crater Lake.**

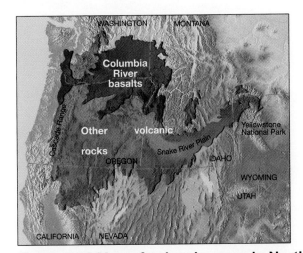

Figure 4.20 **Map of volcanic areas in Northwestern United States, with Columbia River basalts.**

Lutgens/Tarbuck, *Essentials of Geology, 9e*
© 2006 Pearson Prentice Hall, Inc.

Chapter 4

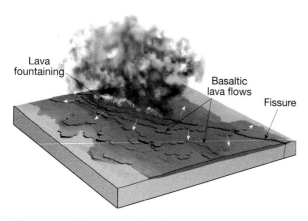

Figure 4.21a **Basaltic fissure eruption.**

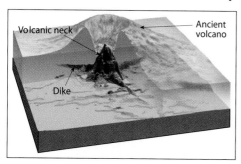

Figure 4.22 **Shiprock, NM, a volcanic neck.**

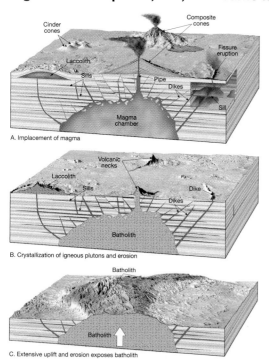

Figure 4.23 **Basic igneous structures.**

Lutgens/Tarbuck, *Essentials of Geology, 9e*
© 2006 Pearson Prentice Hall, Inc.

Volcanoes and Other Igneous Activity   4-5

NOTES:

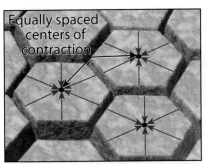
Figure 4.25 Columnar jointing.

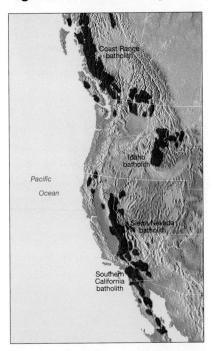
Figure 4.26 Map of west coast showing granitic batholiths.

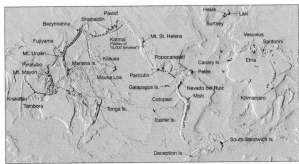
Figure 4.27 Location of some of Earth's major volcanoes.

Lutgens/Tarbuck, *Essentials of Geology*, 9e
© 2006 Pearson Prentice Hall, Inc.

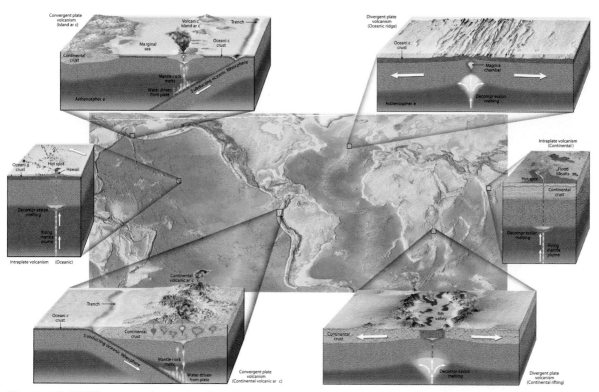

Figure 4.28 Three zones of volcanism.

NOTES:

Volcanoes and Other Igneous Activity 4-7

NOTES:

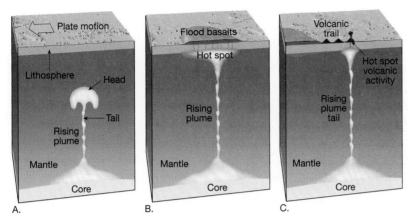

Figure 4.29 Model of mantle plume.

CHAPTER 5 Weathering and Soils

NOTES:

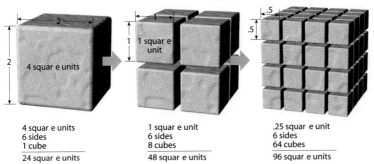

Figure 5.2 **Chemical and mechanical weathering.**

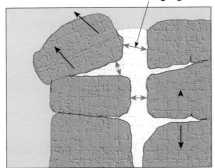

Figure 5.3 **Frost wedging.**

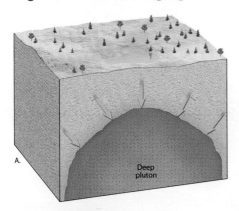

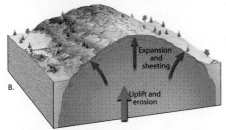

Figure 5.4 **Sheeting.**

Lutgens/Tarbuck, *Essentials of Geology, 9e*
© 2006 Pearson Prentice Hall, Inc.

5-2   Chapter 5

NOTES:

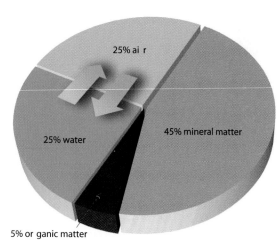

Figure 5.11 Composition of soil.

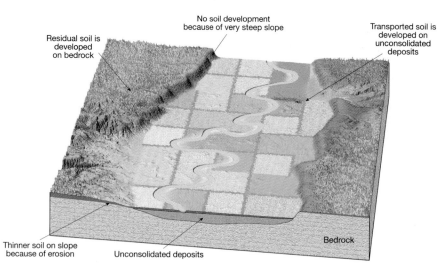

Figure 5.12 Parent material of soils

Weathering and Soils 5-3

NOTES:

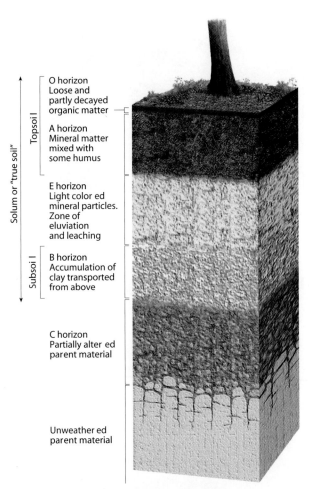

Figure 5.13 Idealized soil profile.

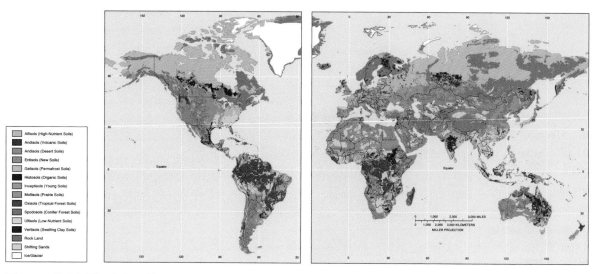

Figure 5.14 World soil map.

NOTES:

CHAPTER 6   Sedimentary Rocks

NOTES:

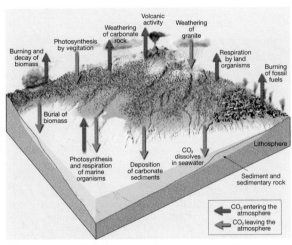

Figure 6.A Diagram of carbon cycle.

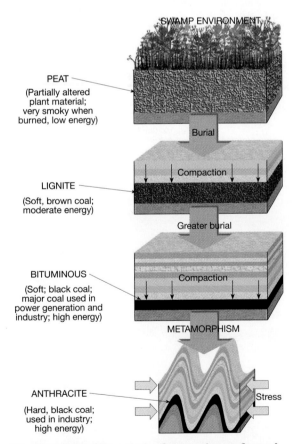

Figure 6.13 Stages in formation of coal.

Detrital Sedimentary Rocks

Clastic Texture (particle size)	Sediment Name	Rock Name
Coarse (over 2 mm)	Gravel (Rounded particles)	Conglomerate
Coarse (over 2 mm)	Gravel (Angular particles)	Breccia
Medium (1/16 to 2 mm)	Sand (If abundant feldspar is present the rock is called **Arkose**)	Sandstone
Fine (1/16 to 1/256 mm)	Mud	Siltstone
Very fine (less than 1/256 mm)	Mud	Shale

Chemical Sedimentary Rocks

Composition	Texture	Rock Name	
Calcite, $CaCO_3$	Nonclastic: Fine to coarse crystalline	Crystalline Limestone	
Calcite, $CaCO_3$	Nonclastic: Fine to coarse crystalline	Travertine	
Calcite, $CaCO_3$	Clastic: Visible shells and shell fragments loosely cemented	Coquina	Biochemical Limestone
Calcite, $CaCO_3$	Clastic: Various size shells and shell fragments cemented with calcite cement	Fossiliferous Limestone	Biochemical Limestone
Calcite, $CaCO_3$	Clastic: Microscopic shells and clay	Chalk	Biochemical Limestone
Quartz, SiO_2	Nonclastic: Very fine crystalline	Chert (light colored) Flint (dark colored)	
Gypsum $CaSO_4 \cdot 2H_2O$	Nonclastic: Fine to coarse crystalline	Rock Gypsum	
Halite, NaCl	Nonclastic: Fine to coarse crystalline	Rock Salt	
Altered plant fragments	Nonclastic: Fine-grained organic matter	Bituminous Coal	

Figure 6.14 Identifying sedimentary rocks.

NOTES:

Sedimentary Rocks 6-3

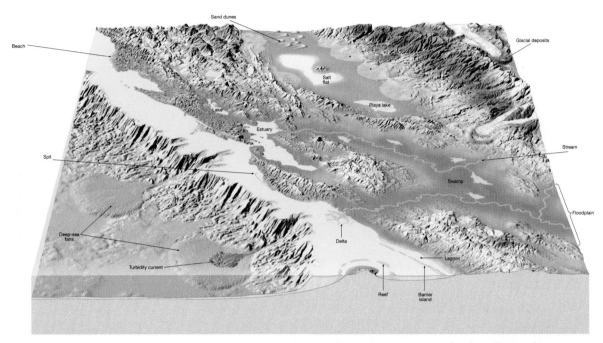

Figure 6.16 Sedimentary environments: dunes, beach, stream, glacier deposits.

NOTES:

6-4   Chapter 6

NOTES:

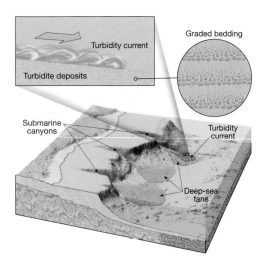

Figure 6.19 Submarine canyons and turbidity currents.

Figure 6.22 U.S. annual per capita use of mineral resources.

Sedimentary Rocks 6-5

NOTES:

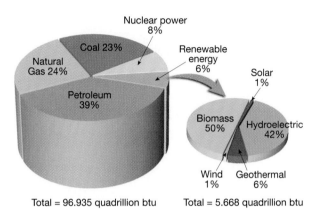
Figure 6.23 U.S. energy consumption, 2001.

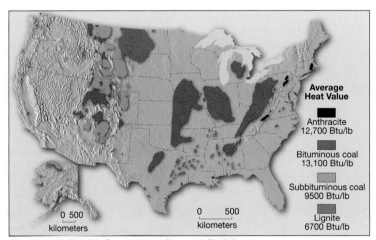
Figure 6.24 U.S. map of coal fields.

CHAPTER 7 Metamorphic Rocks

NOTES:

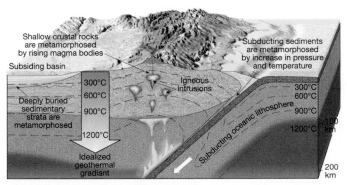

Figure 7.2 Subduction zone.

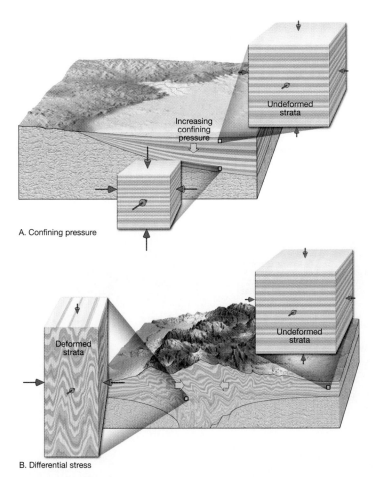

Figure 7.3 Pressure as a metamorphic agent.

NOTES:

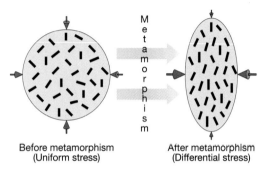

Figure 7.4 Mechanical rotation of platy or elongated mineral grains.

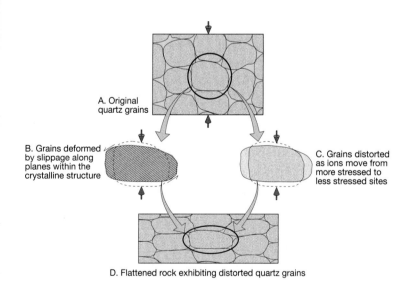

Figure 7.5 Development of preferred orientation of minerals.

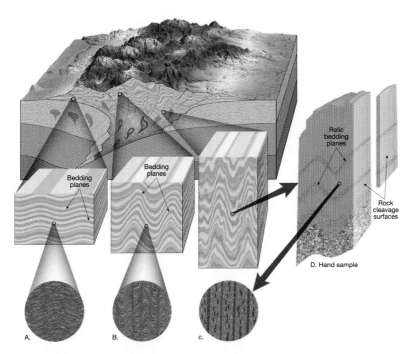

Figure 7.6 Development of one type of rock cleavage.

Rock Name	Texture	Grain Size	Comments	Parent Rock
Slate	Foliated	Very fine	Excellent rock cleavage, smooth dull surfaces	Shale, mudstone, or siltstone
Phyllite	Foliated	Fine	Breaks along wavy surfaces, glossy sheen	Slate
Schist	Foliated	Medium to Coarse	Micaceous minerals dominate, scaly foliation	Phyllite
Gneiss	Foliated	Medium to Coarse	Compositional banding due to segregation of minerals	Schist, granite, or volcanic rocks
Migmatite	Foliated	Medium to Coarse	Banded rock with zones of light-colored crystalline minerals	Gneiss, schist
Mylonite	Weakly Foliated	Fine	When very fine-grained, resembles chert, often breaks into slabs	Any rock type
Metaconglomerate	Weakly Foliated	Coarse-grained	Stretched pebbles with preferred orientation	Quartz-rich conglomerate
Marble	Nonfoliated	Medium to coarse	Interlocking calcite or dolomite grains	Limestone, dolostone
Quartzite	Nonfoliated	Medium to coarse	Fused quartz grains, massive, very hard	Quartz sandstone
Hornfels	Nonfoliated	Fine	Usually, dark massive rock with dull luster	Any rock type
Anthracite	Nonfoliated	Fine	Shiny black rock that may exhibit conchoidal fracture	Bituminous coal
Fault breccia	Nonfoliated	Medium to very coarse	Broken fragments in a haphazard arrangement	Any rock type

(Increasing Metamorphism)

Figure 7.10 Classification of metamorphic rocks.

Lutgens/Tarbuck, *Essentials of Geology, 9e*

© 2006 Pearson Prentice Hall, Inc.

Chapter 7

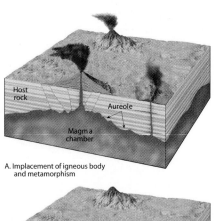

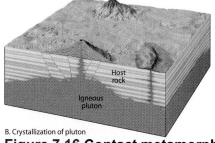

Figure 7.16 Contact metamorphism.

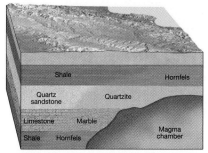

Figure 7.17 Contact metamorphism of shale yields hornfels.

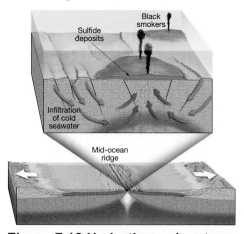

Figure 7.18 Hydrothermal metamorphism along a mid-ocean ridge.

NOTES:

Metamorphic Rocks 7-5

NOTES:

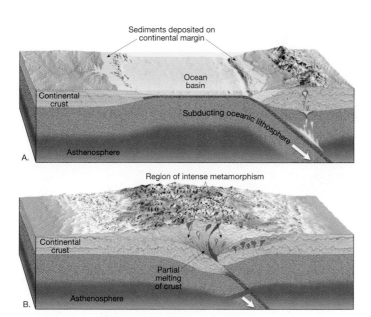

Figure 7.19 Regional metamorphism.

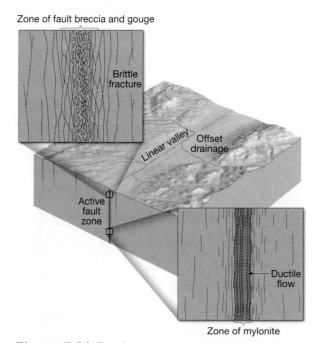

Figure 7.20 Fault zone.

NOTES:

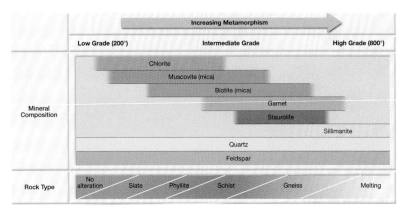

Figure 7.22 Mineral transition results from shale metamorphism.

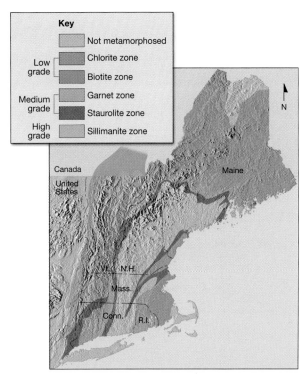

Figure 7.23 Map of U.S. Northeast shows zones of metamorphic intesnities.

CHAPTER 8 Mass Wasting: The Work of Gravity

NOTES:

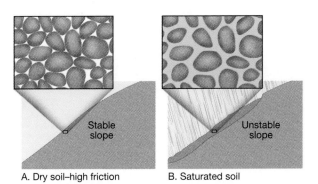

Figure 8.3 **Effect of water on mass wasting.**

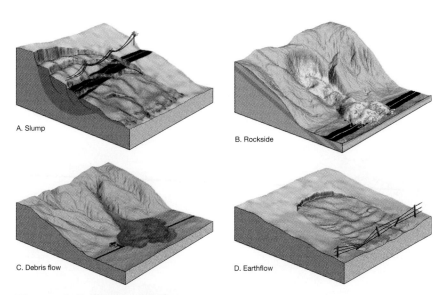

Figure 8.5 Mass wasting.

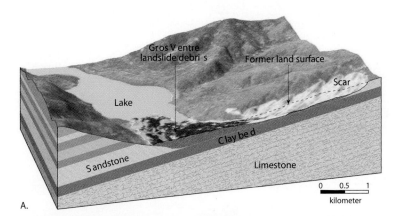

Figure 8.9 Gros Ventre rockslide.

8-2   Chapter 8

NOTES:

Figure 8.A Area of Venezuela affected by disastrous debris flows and flash floods in 1999.

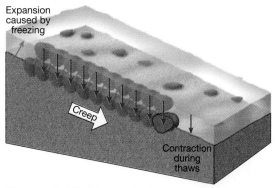

Figure 8.12 Repeated expansion and contraction of surface materials

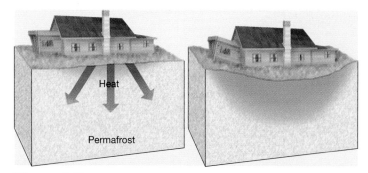

Figure 8.E Subsidence due to permafrost.

CHAPTER 9 Running Water

NOTES:

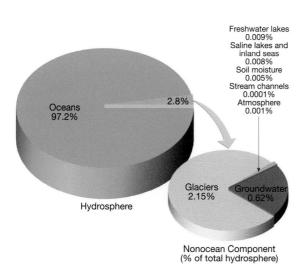

Figure 9.1 Distribution of Earth's water.

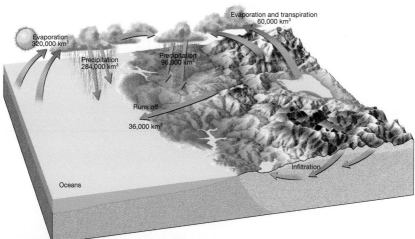

Figure 9.2 Earth's water balance.

Figure 9.3 Drainage basin.

9-2   Chapter 9

NOTES:

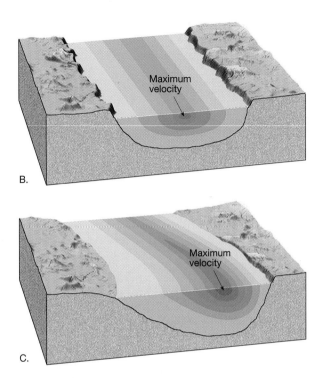

Figure 9.4 b,c Stream velocity.

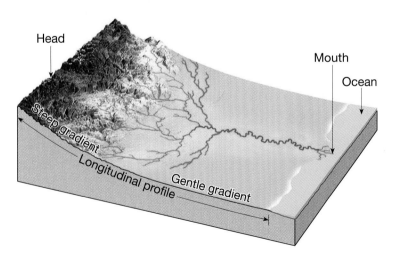

Figure 9.6 A longitudinal profile is a cross-section along the length of a stream.

Running Water 9-3

NOTES:

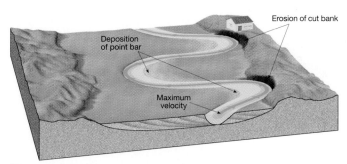

Figure 9.7 Stream meanders.

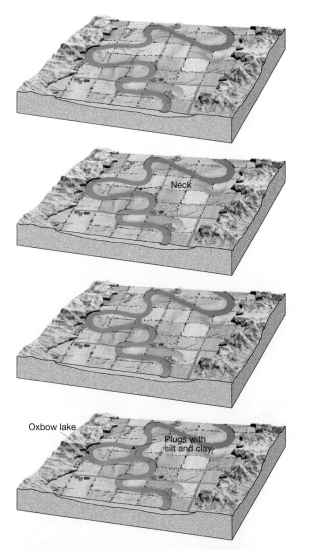

Figure 9.8 Formation of a cutoff and oxbow lake.

NOTES:

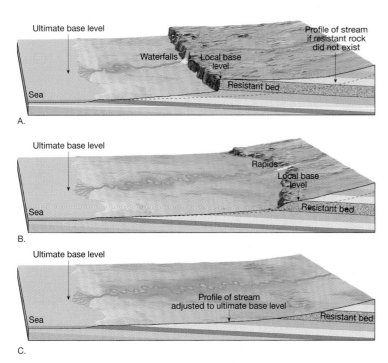

Figure 9.11 **Resistant layer of rock.**

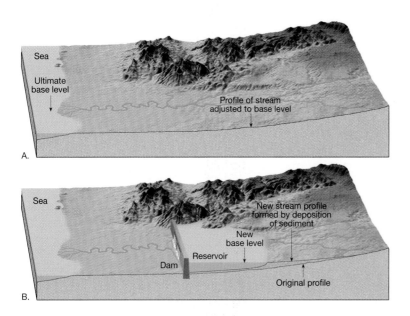

Figure 9.12 **Stream base level is raised.**

NOTES:

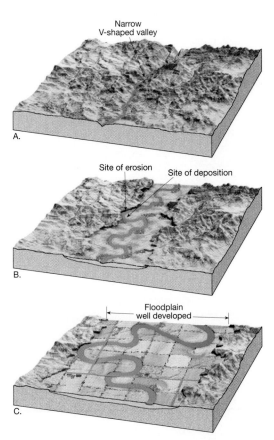

Figure 9.14 **Stream eroding its floodplain.**

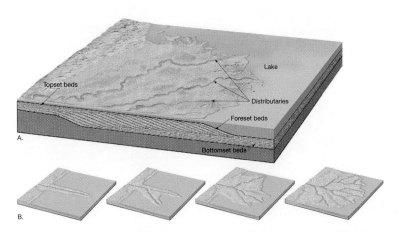

Figure 9.16 a,b **Structure and growth of a simple delta.**

NOTES:

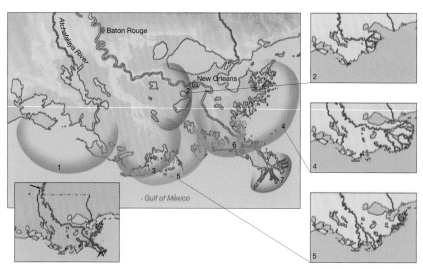

Figure 9.17 Mississippi River deltas.

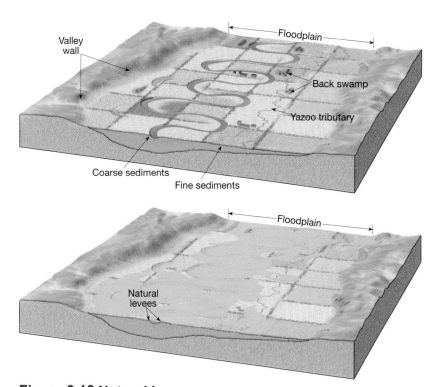

Figure 9.18 Natural levees.

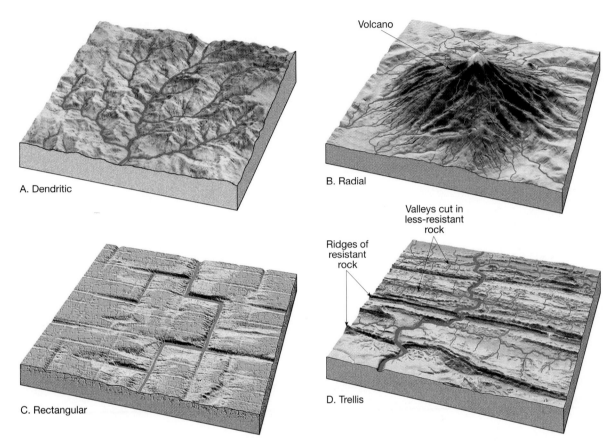

Figure 9.19 Drainage paterns.

NOTES:

CHAPTER 10 Groundwater

NOTES:

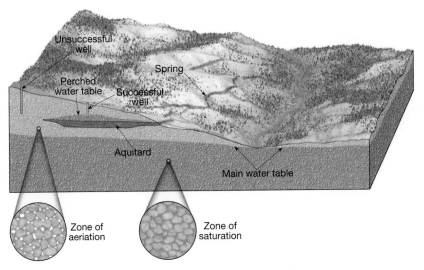

Figure 10.2 Subsurface water landscape features.

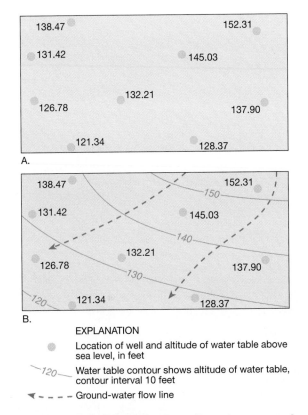

EXPLANATION

- ● Location of well and altitude of water table above sea level, in feet
- —120— Water table contour shows altitude of water table, contour interval 10 feet
- ◄ - - - - Ground-water flow line

Figure 10.3 Distribution of underground water.

Lutgens/Tarbuck, *Essentials of Geology*, 9e
© 2006 Pearson Prentice Hall, Inc.

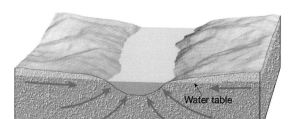

A. Gaining stream

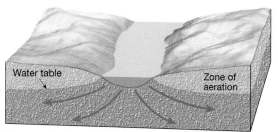

B. Losing stream (connected)

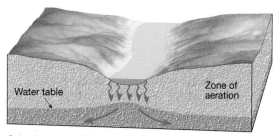

C. Losing stream (disconnected)

Figure 10.4 Gaining stream, losing streams.

NOTES:

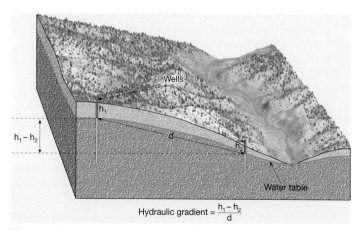

Hydraulic gradient $= \dfrac{h_1 - h_2}{d}$

Figure 10.A Hydraulic gradient.

Groundwater 10-3

NOTES:

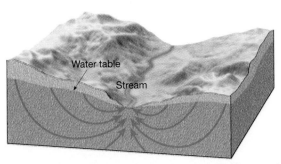

Figure 10.5 Groundwater moves through permeable material.

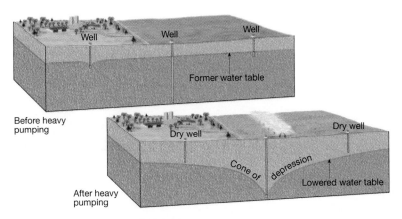

Figure 10.7 Cone of depression.

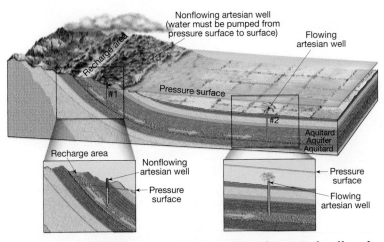

Figure 10.8 Artesian systems occur when an inclined aquifer is surrounded by impermeable beds.

Lutgens/Tarbuck, *Essentials of Geology, 9e*
© 2006 Pearson Prentice Hall, Inc.

NOTES:

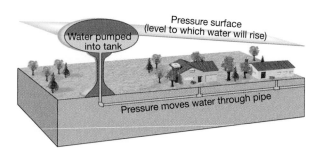

Figure 10.9 City water systems as artesian systems.

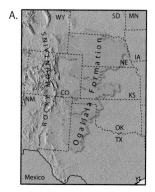

Figure 10.10 Ogallala aquifer map.

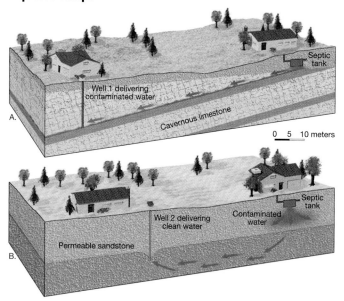

Figure 10.11 Contaminated water.

Groundwater 10-5

NOTES:

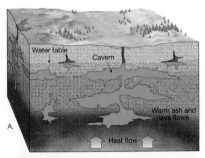
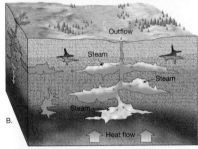
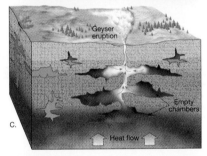

Figure 10.13 Idealized diagrams of a geyser.

NOTES:

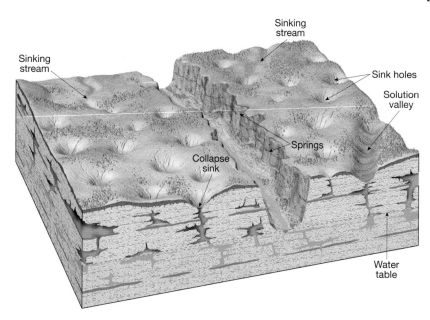

Figure 10.17 **Sinking stream, sinkholes, springs, solution valley, and water table.**

CHAPTER 11 Glaciers and Glaciation

NOTES:

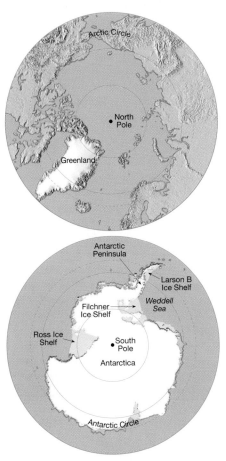

Figure 11.2 Present-day continental ice sheets.

Figure 11.3 Malaspina Glacier.

11-2   Chapter 11

NOTES:

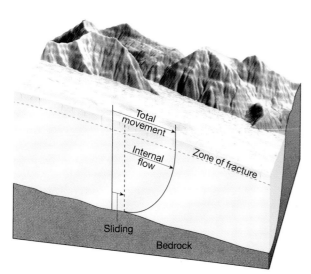

Figure 11.4 Ice movement shown by glacier cross-section.

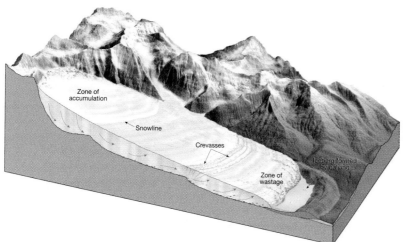

Figure 11.6 Glaciers: zones of wastage and accumulation.

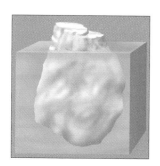

Figure 11.7 Iceberg diagram.

Glaciers and Glaciation 11-3

NOTES:

Figure 11.9 Erosional landforms created by alpine glaciers.

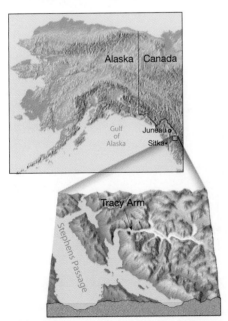

Figure 11.11 Tracy Arm drowned glacial trough.

Lutgens/Tarbuck, *Essentials of Geology, 9e*
© 2006 Pearson Prentice Hall, Inc.

Chapter 11

NOTES:

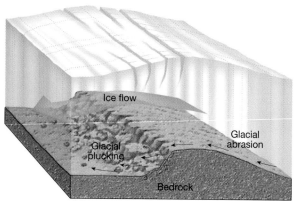
Figure 11.13 Roche moutonnée.

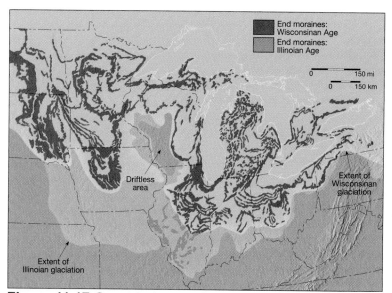
Figure 11.17 Great Lakes region: end moraines.

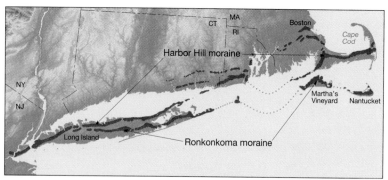
Figure 11.18 New England region: end moraines.

Glaciers and Glaciation 11-5

NOTES:

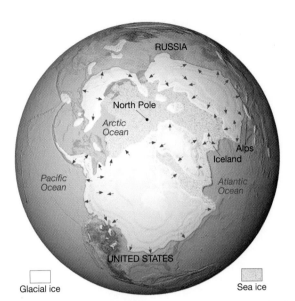

Figure 11.20 Maximum extent of glaciation.

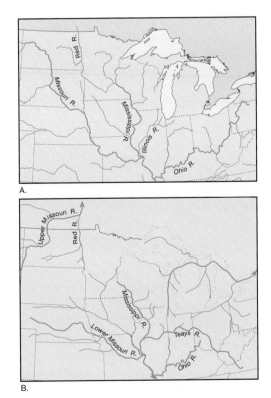

Figure 11.21 Present-day and pre-ice age Great Lakes of U.S.

NOTES:

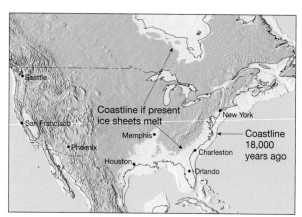

Figure 11.22 Present-day coastline.

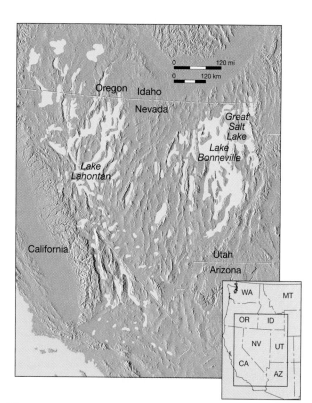

Figure 11.23 Pluvial lakes of western U.S.

Glaciers and Glaciation 11-7

NOTES:

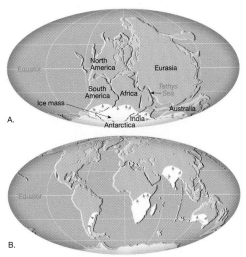

Figure 11.24 Supercontinent Pangaea.

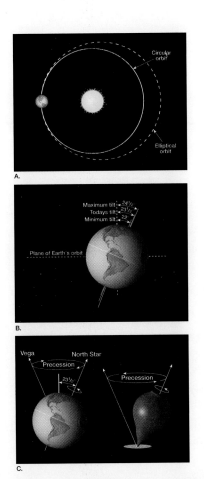

Figure 11.25 Orbital variations.

NOTES:

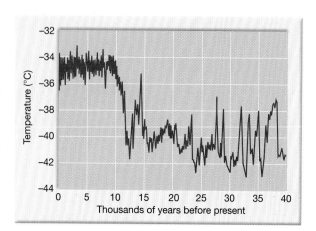

Figure 11.B Temperature flux over last 40,000 years derived from oxygen isotope analysis.

CHAPTER 12 Deserts and Wind

NOTES:

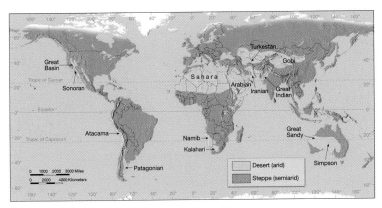

Figure 12.2 Arid and semiarid climates.

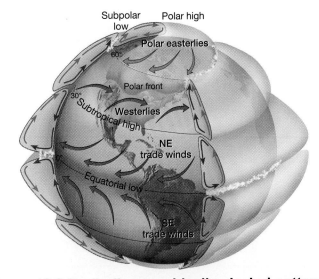

Figure 12.3 Earth diagram, idealized wind patterns.

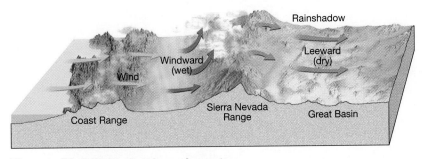

Figure 12.4 Rainshadow desert.

12-2   Chapter 12

NOTES:

Figure 12.A Aral Sea map.

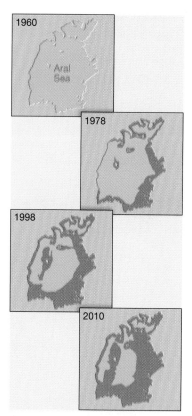

Figure 12.B Shrinking of Aral Sea over time.

Deserts and Wind   12-3

NOTES:

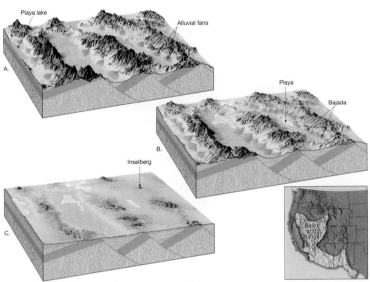
Figure 12.6 Landscape evolution in a mountainous desert.

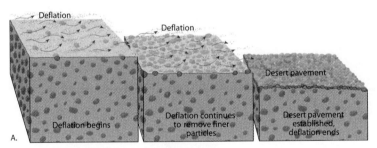
Figure 12.10 Desert pavement.

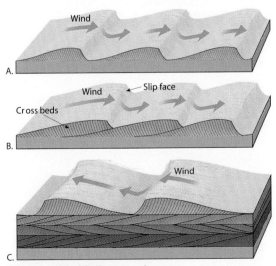
Figure 12.13 Dune shape.

Lutgens/Tarbuck, *Essentials of Geology, 9e*
© 2006 Pearson Prentice Hall, Inc.

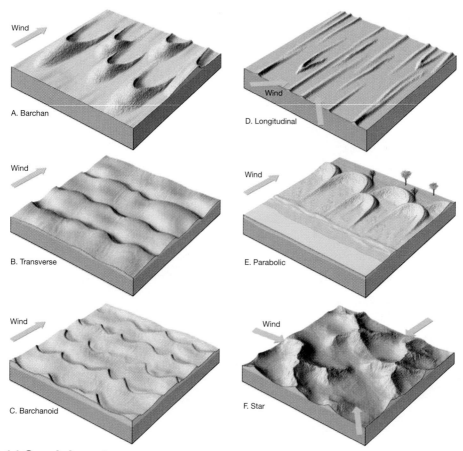

Figure 12.14 Sand dune types.

NOTES:

CHAPTER 13 Shorelines

NOTES:

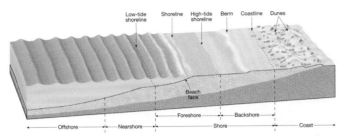

Figure 13.2 Coastal zone parts.

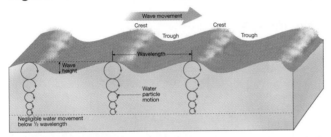

Figure 13.3 Basic wave parts, plus movement of water particles at depth.

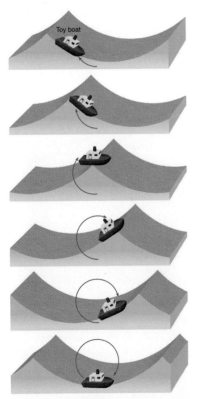

Figure 13.4 Movement of toy boat.

NOTES:

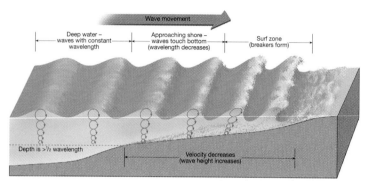

Figure 13.5 Changes that occur when a wave moves onto shore.

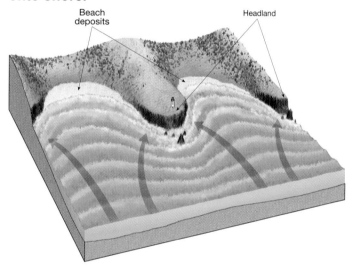

Figure 13.8 Greatest erosional power is concentrated on the headlands.

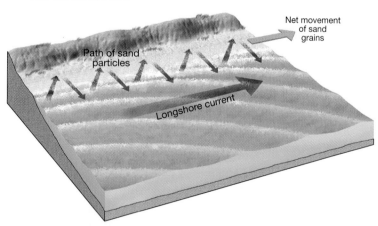

Figure 13.9 Beach drift and longshore currents, obliquely breaking waves.

Lutgens/Tarbuck, *Essentials of Geology, 9e*
© 2006 Pearson Prentice Hall, Inc.

Shorelines 13-3

NOTES:

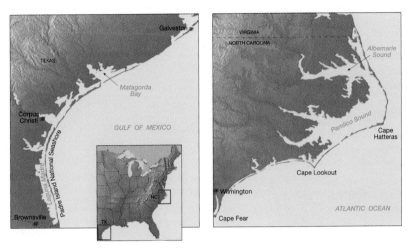

Figure 13.13 Barrier islands, shore of Texas.

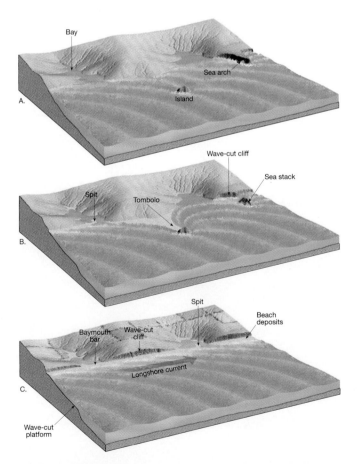

Figure 13.14 Initially irregular coastline remains relatively stable.

NOTES:

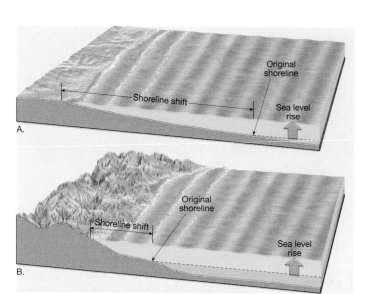

Figure 13.B a,b Shoreline slope and sea level change.

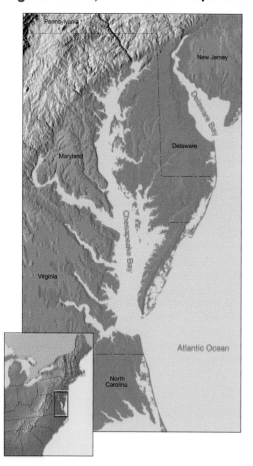

Figure 13.19 Estuaries along East Coast of U.S.

Shorelines 13-5

NOTES:

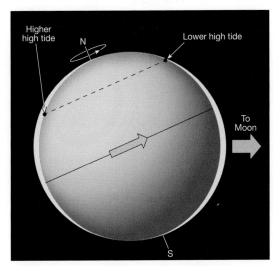
Figure 13.21 Earth model, tides.

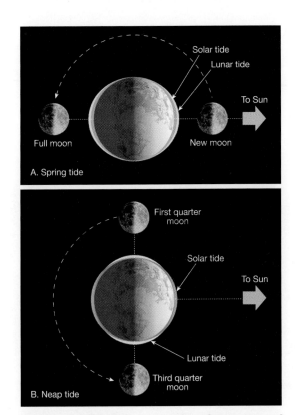
Figure 13.22 Relationship of moon and sun to Earth during spring and neap tides.

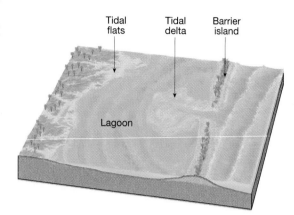

Figure 13.23 Tidal delta.

NOTES:

CHAPTER 14 Earthquakes and Earth's Interior

NOTES:

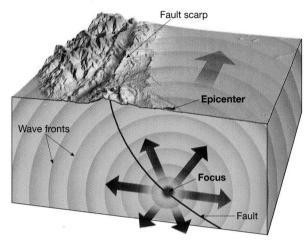

Figure 14.2 Focus of earthquakes.

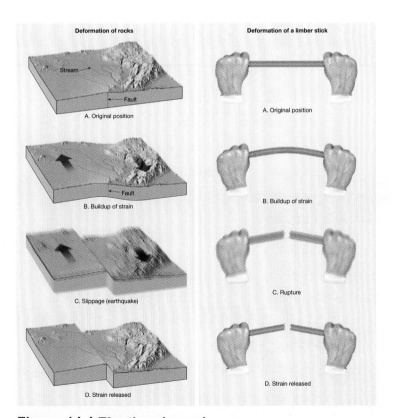

Figure 14.4 Elastic rebound.

NOTES:

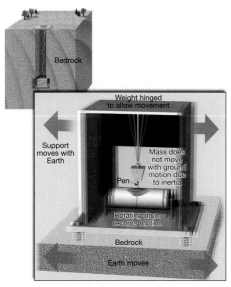

Figure 14.5 Principle of the seismograph.

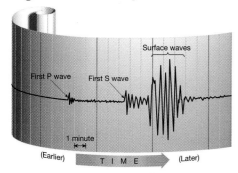

Figure 14.6 Typical seismic record.

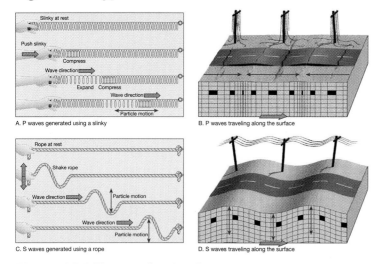

Figure 14.7 Types of seismic waves.

Earthquakes and Earth's Interior 14-3

NOTES:

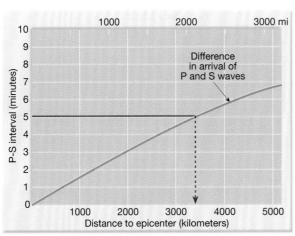

Figure 14.8 Time graph used to determine distance to epicenter.

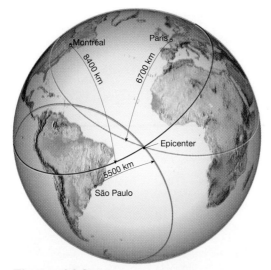

Figure 14.9 Earthquake epicenter.

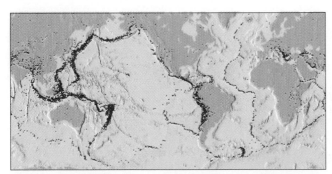

Figure 14.10 World distribution of earthquakes.

NOTES:

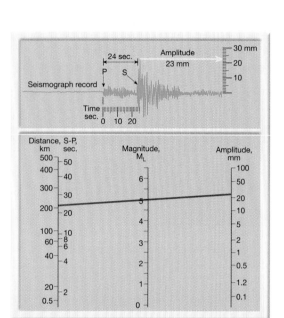

Figure 14.11 Richter magnitude determination using graph.

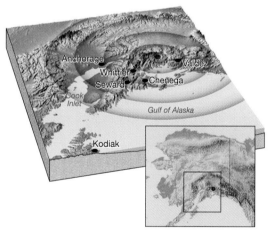

Figure 14.12 Region most affected by Good Friday earthquake 1964.

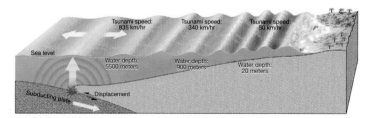

Figure 14.16 Schematic of tsunami.

Earthquakes and Earth's Interior   14-5

NOTES:

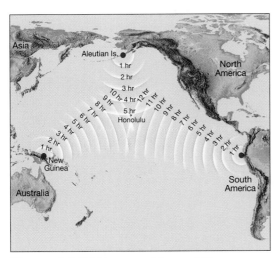

Figure 14.17 Tsunami travel times.

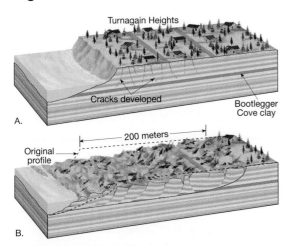

Figure 14.19 a,b Turnagain Heights slide.

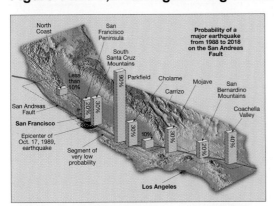

Figure 14.21 Probability of a major earthquake 1988-2018 on the San Andreas fault.

Lutgens/Tarbuck, *Essentials of Geology, 9e*
© 2006 Pearson Prentice Hall, Inc.

14-6   Chapter 14

NOTES:

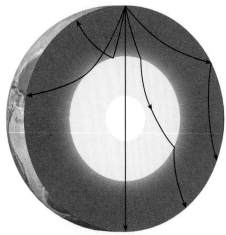

Figure 14.22 Schematic of Earth showing paths of seismic rays.

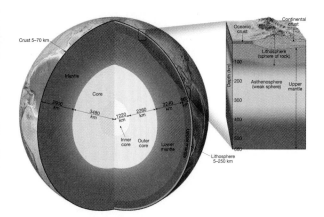

Figure 14.23 Cross-sectional view of Earth.

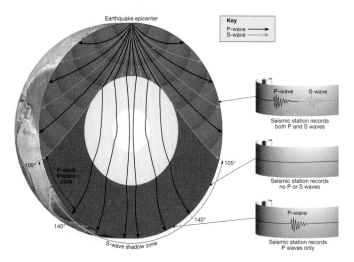

Figure 14.24 Cross-section showing P and S wave paths.

Lutgens/Tarbuck, *Essentials of Geology, 9e*
© 2006 Pearson Prentice Hall, Inc.

CHAPTER 15 Plate Tectonics: A Scientific Theory Unfolds

NOTES:

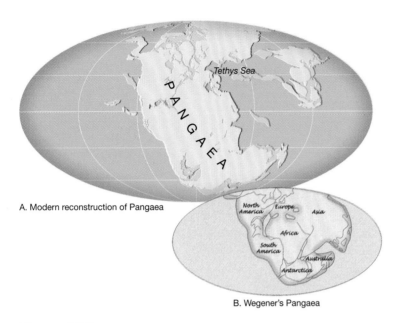

Figure 15.2 Pangaea map.

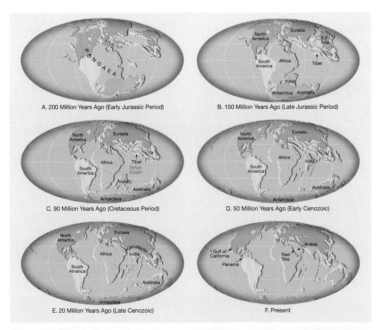

Figure 15.A Several views of the breakup of Pangaea.

15-2   Chapter 15

NOTES:

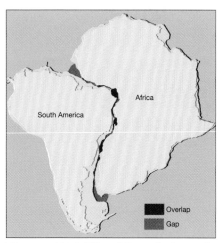
Figure 15.3 Fit of South America and Africa.

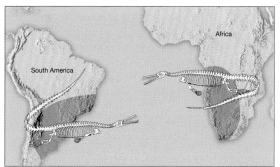
Figure 15.4 Fossils of Mesosaurus have been found on both sides of the South Atlantic and nowhere else.

Figure 15.6 Matching mountain ranges across the North Atlantic.

Lutgens/Tarbuck, *Essentials of Geology, 9e*
© 2006 Pearson Prentice Hall, Inc.

Plate Tectonics: A Scientific Theory Unfolds 15-3

NOTES:

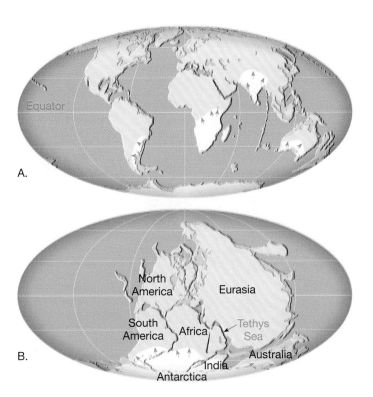

Figure 15.7 Paleoclimatic evidence for continental drift.

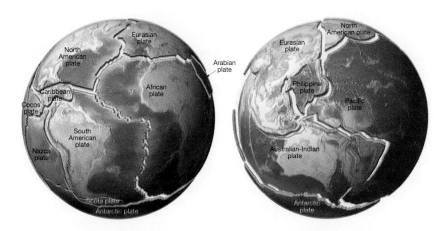

Figure 15.8 Some of Earth's lithospheric plates.

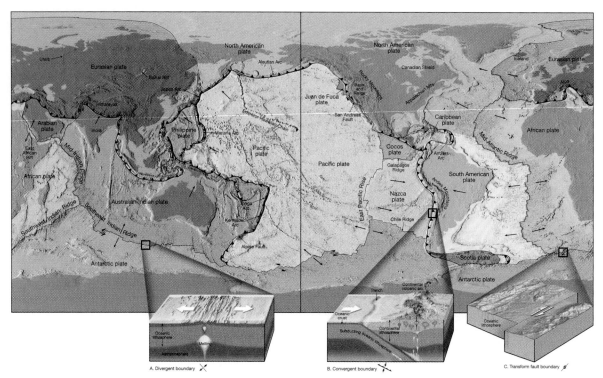

Figure 15.9 **World map shows tectonic plates.**

NOTES:

Plate Tectonics: A Scientific Theory Unfolds 15-5

NOTES:

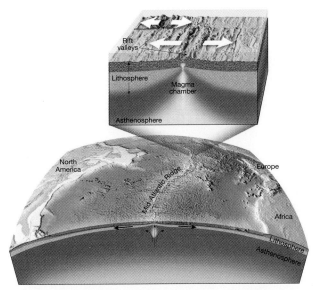

Figure 15.10 Divergent plate boundaries located along the crests of oceanic ridges.

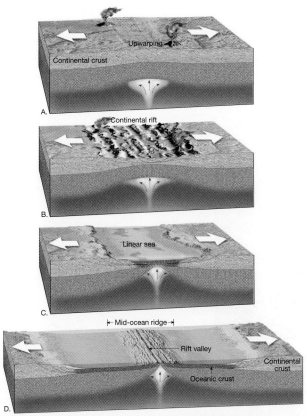

Figure 15.11 Formation of ocean basin.

15-6   Chapter 15

NOTES:

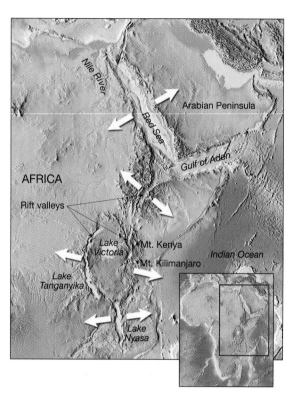

Figure 15.12 East African rift valleys and associated features.

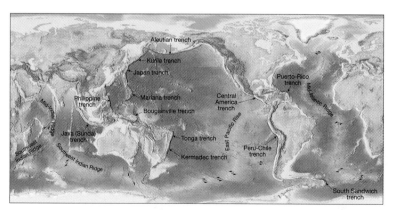

Figure 15.13 Distribution of oceanic trenches, ridge system, and transform faults.

Plate Tectonics: A Scientific Theory Unfolds   15-7

NOTES:

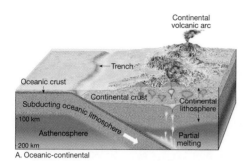
A. Oceanic-continental

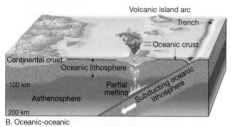
B. Oceanic-oceanic

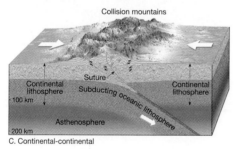
C. Continental-continental

Figure 15.14 Zones of plate convergence.

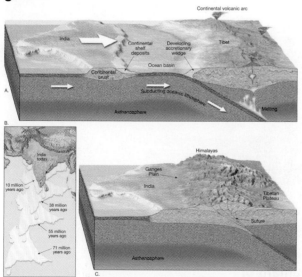

Figure 15.15 Movement of Indian subcontinent over time.

Lutgens/Tarbuck, *Essentials of Geology, 9e*
© 2006 Pearson Prentice Hall, Inc.

NOTES:

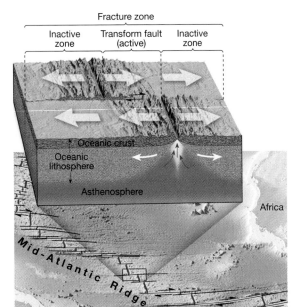

Figure 15.16 Transform faults offsetting segments of a divergent boundary.

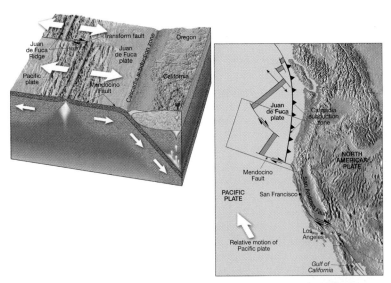

Figure 15.17 Mendocino fault permits movement of seafloor at Juan de Fuca ridge.

Plate Tectonics: A Scientific Theory Unfolds

NOTES:

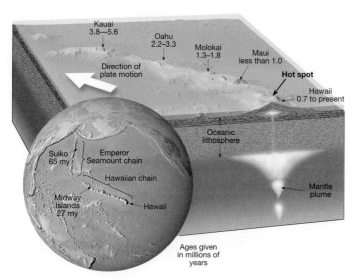

Figure 15.18 Island chain from Hawaii to Aleutians resulted from plate movement over hot spot.

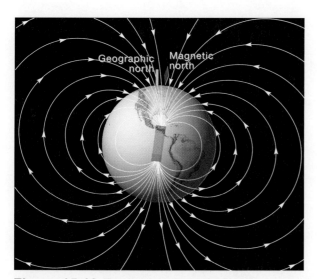

Figure 15.19 Earth's magnetic field.

NOTES:

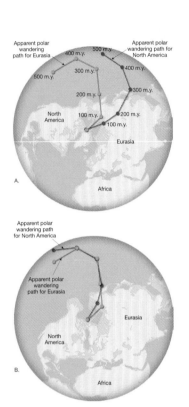

Figure 15.20 **Apparent polar-wandering paths.**

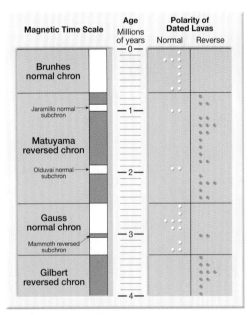

Figure 15.21 **Time scale of Earth's magnetic field.**

NOTES:

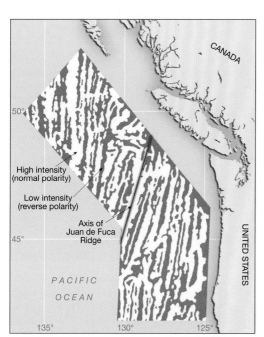

Figure 15.22 Patterns of alternating stripes of high and low intensity magnetism.

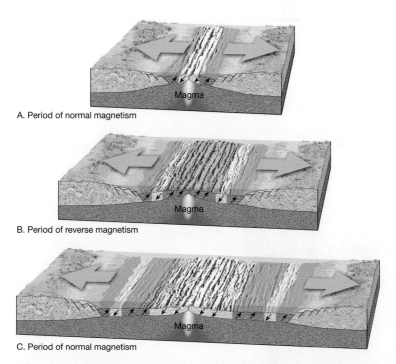

Figure 15.23 New basalt is magnetized according to Earth's exisiting magnetic fields.

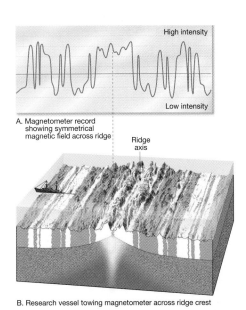

Figure 15.24 **Ocean floor as a magnetic tape recorder.**

Figure 15.25 **Direction and rates of plate motion.**

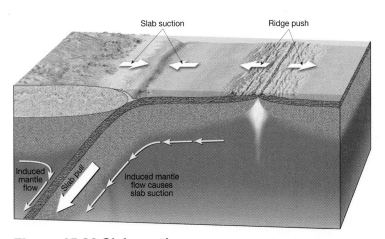

Figure 15.26 **Slab suction.**

NOTES:

NOTES:

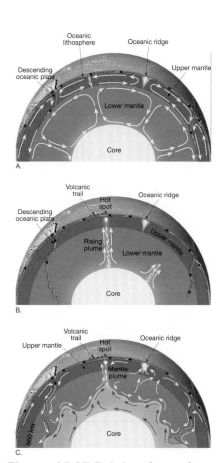

Figure 15.27 Driving force for plate tectonics.

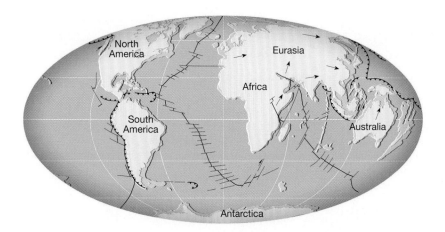

Figure 15.28 Speculative world map, 50 million years in the future.

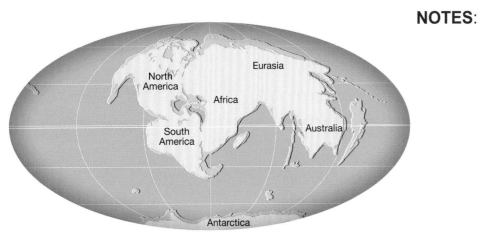

Figure 15.29 Speculative world map, 250 million years in the future.

NOTES:

CHAPTER 16 Origin and Evolution of the Ocean Floor

NOTES:

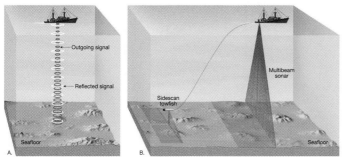
Figure 16.2 Echo sounder.

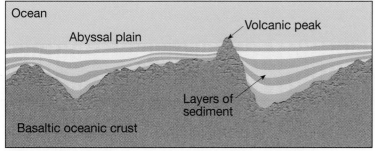
Figure 16.3 Sketch of a seismic cross-section.

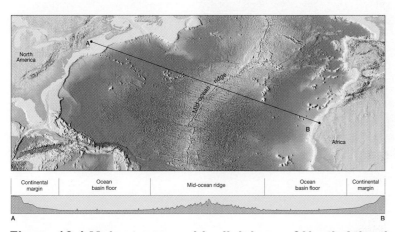
Figure 16.4 Major topographic divisions of North Atlantic.

NOTES:

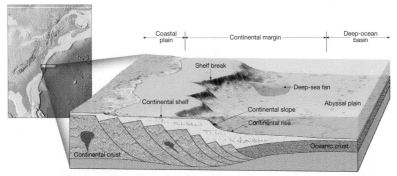

Figure 16.5 Continental margins.

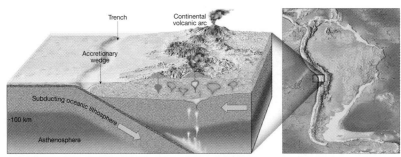

Figure 16.6 Active continental margins.

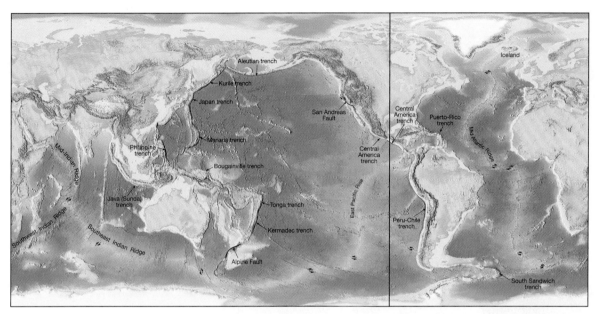

Figure 16.7 **Oceanic trenches, ridge systems, fracture zones, and transform faults.**

NOTES:

NOTES:

Figure 16.B Formation of coral atoll.

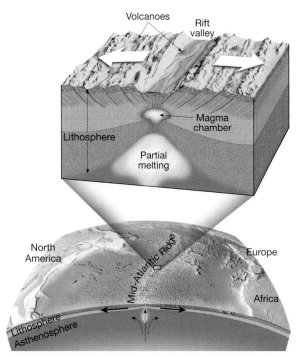

Figure 16.9 Undersea ridges.

Figure 16.10 Divergent plate boundaries located along the crests of oceanic ridges.

Origin and Evolution of the Ocean Floor 16-5

NOTES:

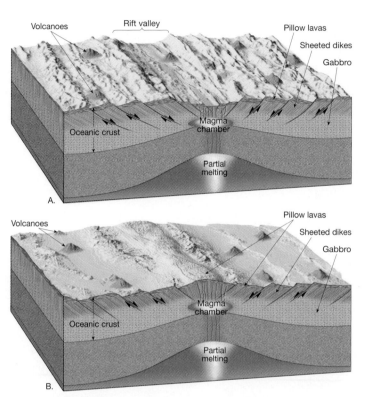

Figure 16.11 Block diagram: undersea volcano, rift valley, and lavas.

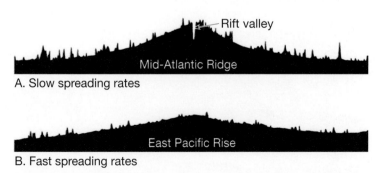

Figure 16.12 Cross section profile of Mid-Atlantic Ridge and East Pacific Rise.

NOTES:

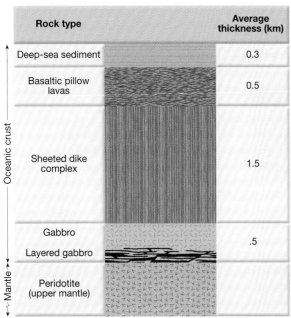

Figure 16.13 Rock types, oceanic crust.

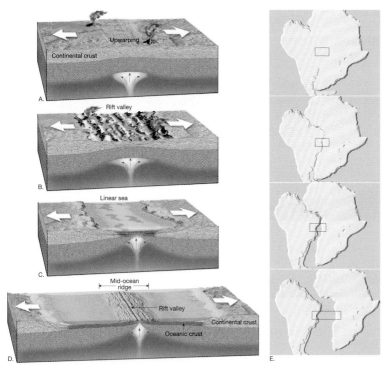

Figure 16.15 Formation of ocean basin.

Origin and Evolution of the Ocean Floor 16-7

NOTES:

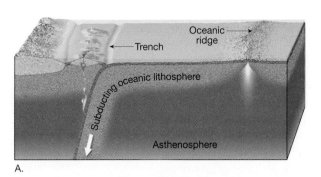
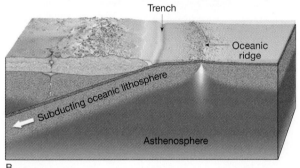

Figure 16.16 Angle of oceanic lithosphere descent into asthenosphere depends on density.

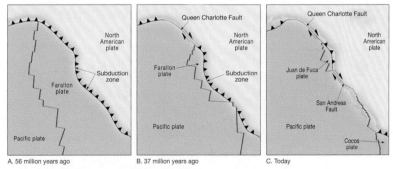

Figure 16.17 Demise of Farallon plate, which once ran along western edge of Americas.

A. 600 million years ago

B. 510 million years ago

C. 430 million years ago

D. 230 million years ago

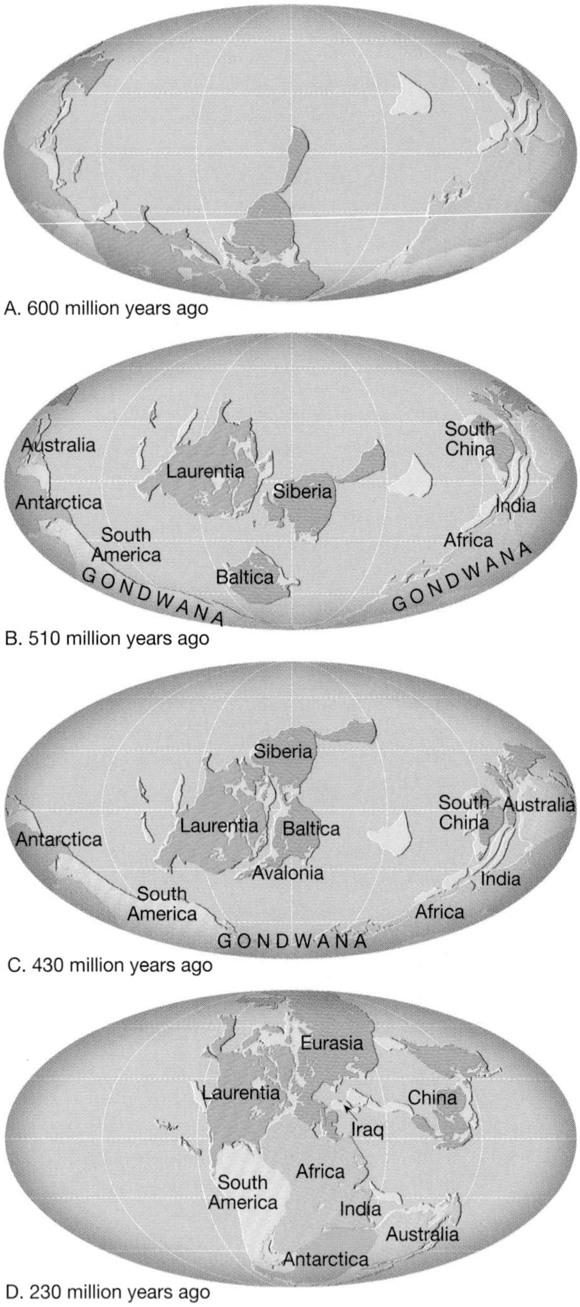

Figure 16.18 Breakup of supercontinent Rodinia and reassembly of fragments into Pangaea.

NOTES:

CHAPTER 17 Crustal Deformation and Mountain Building

NOTES:

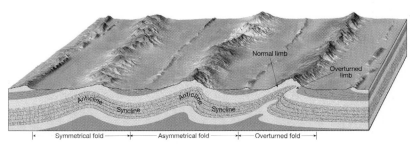

Figure 17.3 **Principal types of folded strata.**

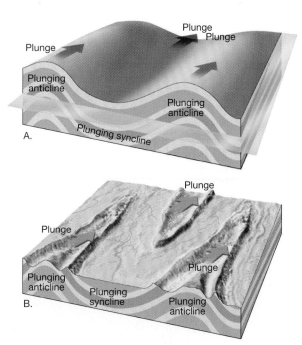

Figure 17.4 **Plunging folds.**

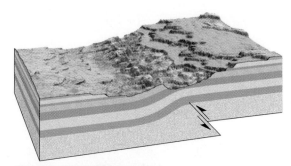

Figure 17.5 **Monocline.**

NOTES:

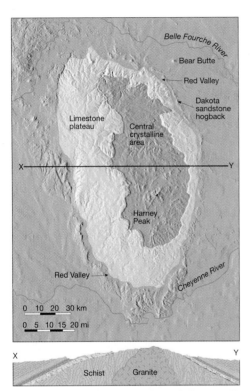

Figure 17.6 Black Hills of South Dakota.

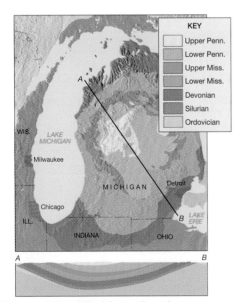

Figure 17.7 Michigan basin, bedrock geology map.

Crustal Deformation and Mountain Building 17-3

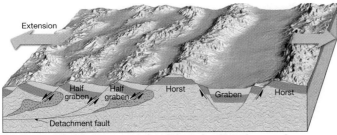

Figure 17.9 **Four types of faults.**

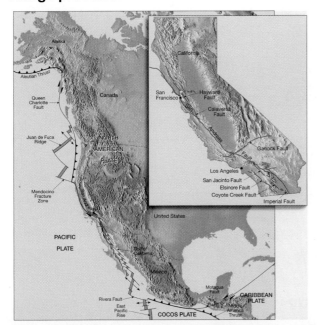

Figure 17.10 **Normal faulting in Basin and Range province.**

Figure 17.A **Extent of San Andreas fault system.**

NOTES:

NOTES:

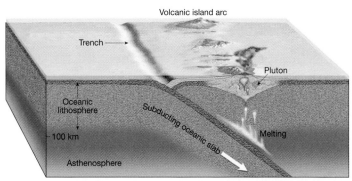

Figure 17.13 **Development of a volcanic island arc.**

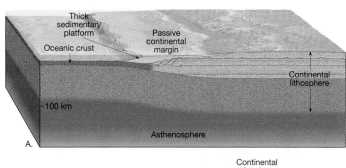

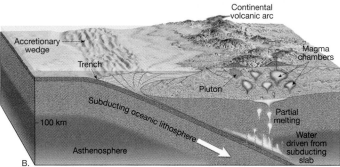

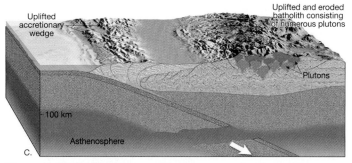

Figure 17.14 **Orogenesis along an Andean-type subduction zone.**

Crustal Deformation and Mountain Building 17-5

NOTES:

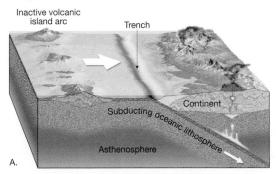

Figure 17.15 Oceanic plateaus and submerged crustal fragments.

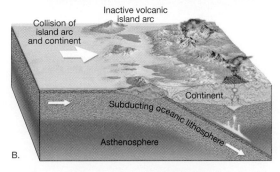

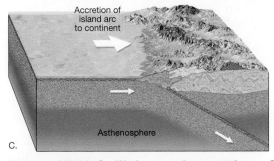

Figure 17.16 Collision and accretion of an arc to a continental margin.

NOTES:

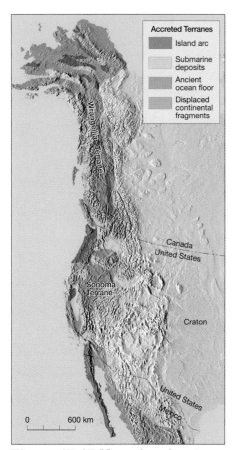

Figure 17.17 **Map showing terranes added to western North America.**

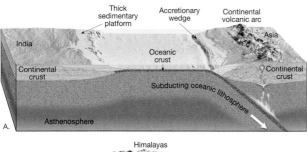

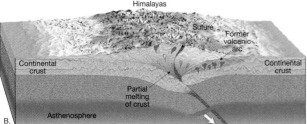

Figure 17.18 **Collision of India and Eurasian plate producing Himalaya Mountains.**

NOTES:

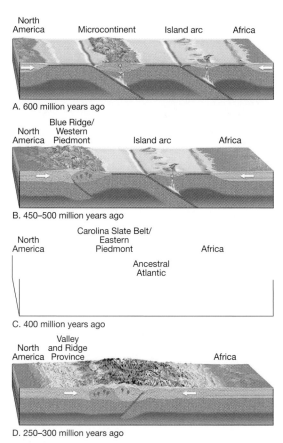

Figure 17.19 Development of southern Appalachians.

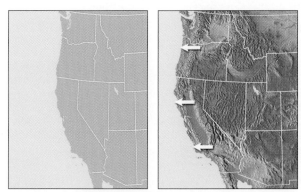

Figure 17.20 Basin and range.

NOTES:

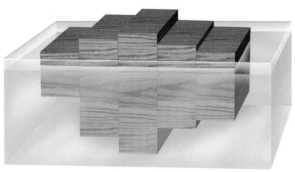

Figure 17.21 **Wooden blocks of different thickness float in water.**

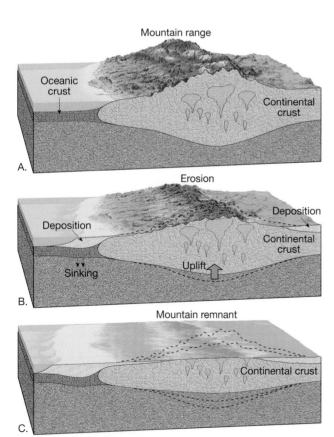

Figure 17.22 **Combined effect of erosion and isostatic adjustment.**

Crustal Deformation and Mountain Building 17-9

NOTES:

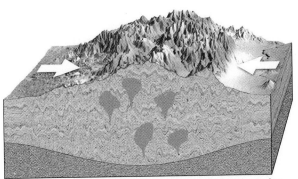

A. Horizontal compressional forces dominate causing shortening and thickening of the crust

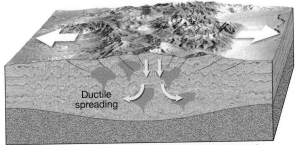

B. Gravitational forces dominate resulting in stretching and thinning of the crust

Figure 17.23 Gravitational collapse.

CHAPTER 18 Geologic Time

NOTES:

B.
Figure 18.2 Grand Canyon.

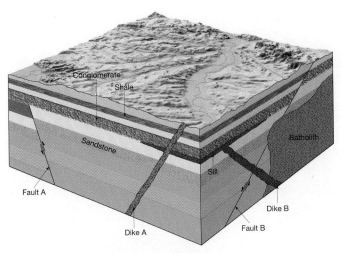

Figure 18.4 Relative dating.

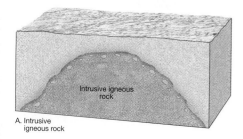

A. Intrusive igneous rock

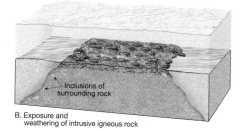

B. Exposure and weathering of intrusive igneous rock

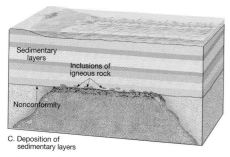

C. Deposition of sedimentary layers

Figure 18.5 Two ways that inclusions can form as well as a type of unconformity.

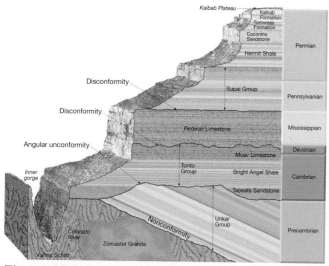

Figure 18.6 Grand Canyon cross-section.

Geologic Time 18-3

NOTES:

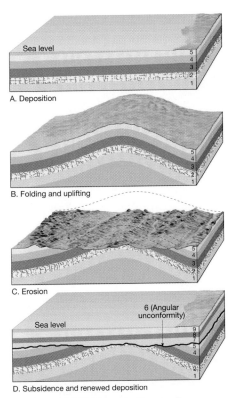

Figure 18.7 **Formation of an angular unconformity.**

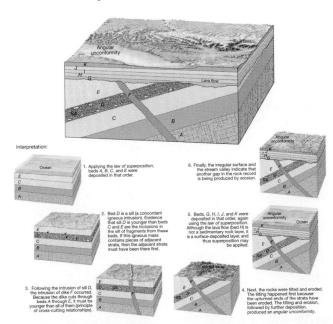

Figure 18.8 **Geologic cross-section of a hypothetical region.**

Lutgens/Tarbuck, *Essentials of Geology, 9e*
© 2006 Pearson Prentice Hall, Inc.

NOTES:

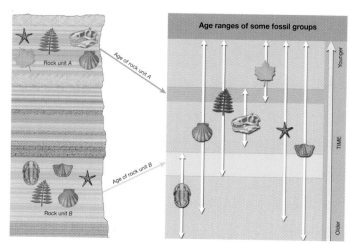

Figure 18.11 **Overlapping ranges of fossils.**

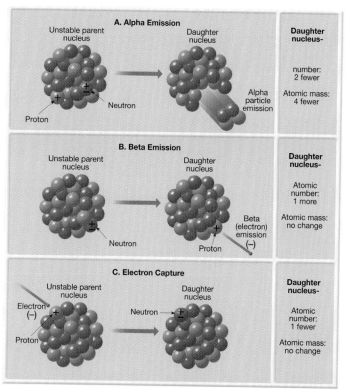

Figure 18.12 **Common types of radioactive decay.**

Geologic Time   18-5

NOTES:

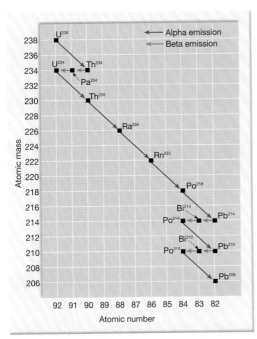

Figure 18.13 **Most common isotope of uranium.**

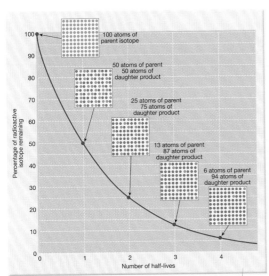

Figure 18.14 **Radioactive decay curve.**

NOTES:

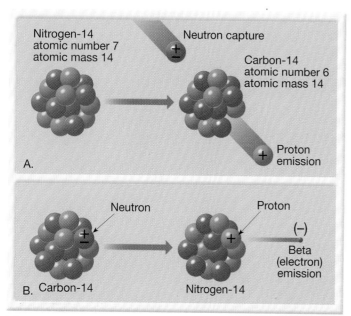

Figure 18.15 **Production and decay of carbon-14.**

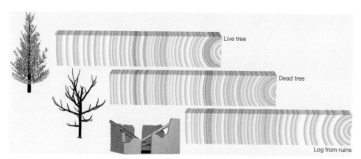

Figure 18.B **Cross dating: dendochronology.**

Geologic Time 18-7

NOTES:

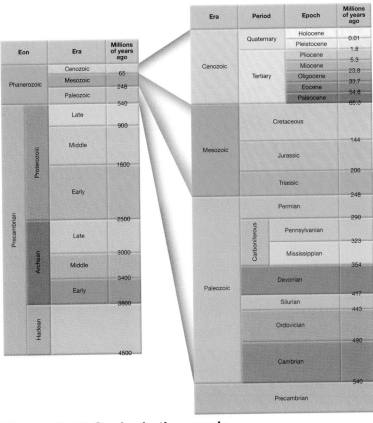

Figure 18.16 Geologic time scale.

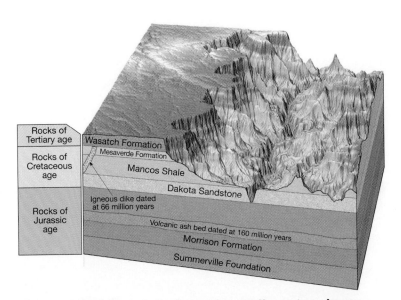

Figure 18.17 Absolute dates for sedimentary layers.

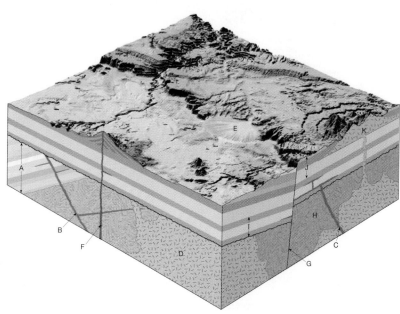

Figure 18.18 Block diagram of hypothetical area in American Southwest.

NOTES:

CHAPTER 19 Earth History: A Brief Summary

NOTES:

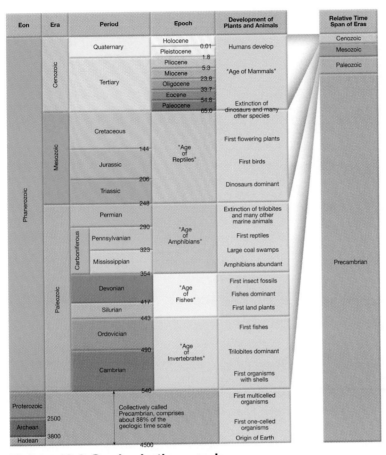

Figure 19.2 Geologic time scale.

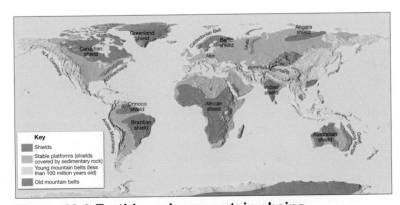

Figure 19.3 Earth's major mountain chains.

Figure 19.4 **Ancient globe.**

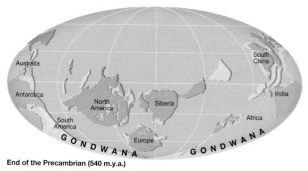
Figure 19.5 **End of Precambrian.**

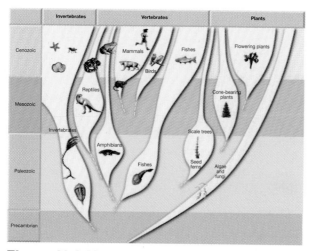
Figure 19.8 **Major groups of organisms.**

NOTES:

NOTES:

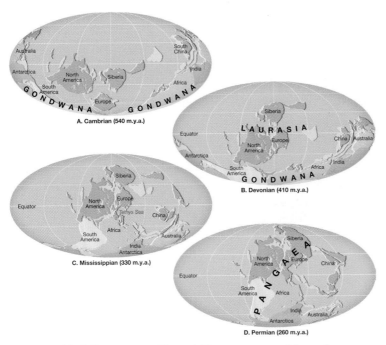

Figure 19.9 Supercontinent Pangaea and breakup.

Figure 19.11 "Plate-skinned" fish.

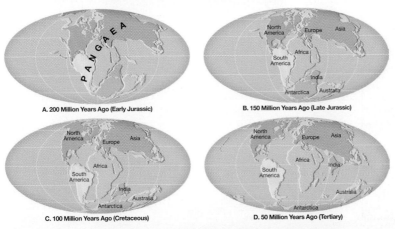

Figure 19.13 Breakup of supercontinent Pangaea.

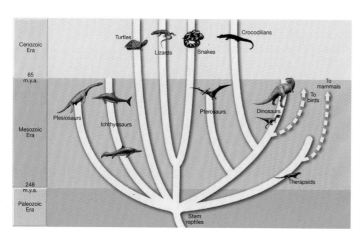

Figure 19.14 **Adaptive radiation of major groups of reptiles.**

Figure 19.15 **Fossils of Pteranodon.**

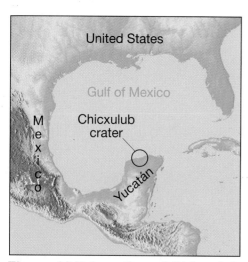

Figure 19.B **Location of Chicxulub crater.**

NOTES:

Earth History: A Brief Summary 19-5

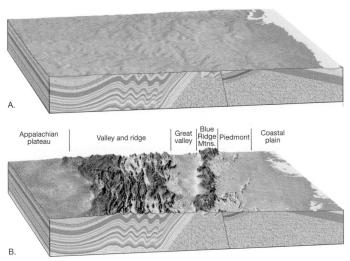

Figure 19.16 Formation of modern Appalachian Mountains.

NOTES:

Appendices

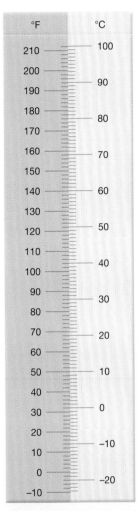

Figure A.1 Temperature scales.

NOTES:

NOTES:

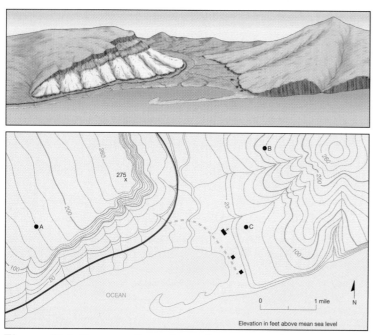

Figure B.1 Perspective view of area; contour map of same area.

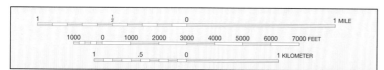

Figure B.2 Graphic scale.

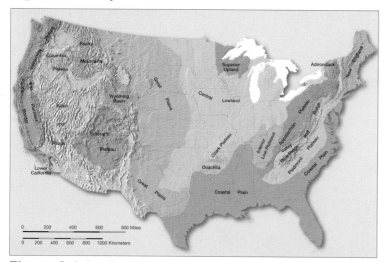

Figure C.1 Outline U.S. map: Major landform regions.

Lutgens/Tarbuck, *Essentials of Geology, 9e*
© 2006 Pearson Prentice Hall, Inc.

Appendices   A-3

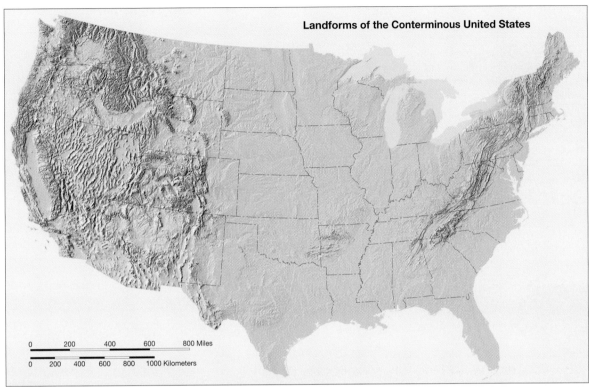

Figure C.2 Landforms of the conterminous U.S.

NOTES: